The Playfair Collection

and the Teaching of Chemistry at the University of Edinburgh 1713–1858

R G W Anderson

The Royal Scottish Museum
Edinburgh 1978

©Crown copyright 1978
First published 1978

ISBN 0 900733 16 0

CONTENTS

	Page
List of Illustrations	v
Foreword	vii
Preface	1
Chapter 1: Early Chemistry Teaching in Edinburgh: James Crawford, Andrew Plummer and William Cullen	3
Chapter 2: Joseph Black	19
Chapter 3: Thomas Charles Hope	35
Chapter 4: William Gregory	47
Chapter 5: Playfair Collection: Acquisition and Provenance	57
CATALOGUE OF THE PLAYFAIR COLLECTION	63
Contents of the Catalogue Section	65
Instruments and Apparatus in the Playfair Collection	67
Appendix 1: Instruments and Apparatus no longer extant	151
Appendix 2: Transcript of Original Entries of 1858 in Register Book	155
Appendix 3: Instruments and Apparatus associated with Edinburgh Professors in other Collections	157
Select Bibliography	161
Index of Personal Names	169

CONTENTS

	Page
List of Organisations	v
2. Edge...	
3. The...	vii
4. Adam...	
5. Lizar...	
6. William...	1
7. Joseph...	
Chapter 1: Early Chemistry Teaching in Edinburgh: James Crawford, Andrew Plummer	
9. Willis and William Cullen	3
10. Lyon...	
11. Hast... Joseph Black	
12. Detail...	19
13. Med...	
Chapter 2: Thomas Charles Hope	35
14. Visit...	
15. Paul...	
16. Balth... William Gregory	47
17. Doub...	
18. Mass... Playfair Collection: Acquisition and Provenance	57
19. Wire...	
20. Pyro...	
CATALOGUE OF THE PLAYFAIR COLLECTION	63
21. Mode...	
22. Air T...	
23. Contents of the Catalogue Section	65
24. Volta...	
25. Instruments and Apparatus in the Playfair Collection	67
26. Volta...	
27. Inst...	
Appendix 1: Instruments and Apparatus no longer extant	151
28. Aud...	
29. Lamp...	
30. Appendix 2: Transcript of Original Entries of 1858 in Register Book	155
31. Ret...	
32. Appendix 3: Instruments and Apparatus associated with Edinburgh Professors	
33. Adjus... in other Collections	157
34. Mult...	
35. Still...	
Select Bibliography	161
36. Blast...	
37. Wate...	
Index of Personal Names	169
39. Bead...	
40. Patte...	
41. Cone...	
42. Theat...	
43. Iron...	
44. Absor...	
45. Bottl...	
46. Grad...	
47. Stopp...	
48. Cucur...	
49. Bottle...	
50. Flask...	

iii

LIST OF ILLUSTRATIONS

		Page
1.	George Wilson, ca. 1850	viii
2.	Edgar's plan of Edinburgh, 1742	6
3.	The College buildings prior to 1789	6
4.	Adam's plan of the University, 1791	7
5.	Lizar's plan of the College	7
6.	William Cullen, 1787	12
7.	Joseph Black, 1787	25
8.	Thomas Charles Hope, 1817	36
9.	William Gregory, ca. 1845	51
10.	Lyon Playfair, 1852	59
11.	Hauksbee-type air pump	67
12.	Detail of air pump	67
13.	Mechanism of air pump	68
14.	Vream's air pump mechanism, 1744	68
15.	Pneumatic trough	71
16.	Balance, with lead block	73
17.	Double bellows	77
18.	Magellan's portable laboratory, 1788	77
19.	Wire cage	80
20.	Pyrometer	82
21.	Model of spiral thermometer	84
22.	Air Thermoscope	86
23.	Differential thermoscopes	88
24.	Voltaic pile	91
25.	Sturgeon cell	94
26.	Voltameter	96
27.	Instantaneous light boxes	99
28.	Argand lamp	101
29.	Lamp furnace	103
30.	Perceval's lamp furnace, 1791	103
31.	Retort stand, boss and holders	105
32.	Heating bath	107
33.	Adjustable table support	108
34.	Multi-purpose apparatus support	109
35.	Still boiler	110
36.	Black's still, 1786	110
37.	Water-cooled condensers	112
38.	Iron bottle	115
39.	Beddoes' and Watt's pneumatic apparatus, 1796	115
40.	Pattern for beehive shelves	118
41.	Concave reflectors	120
42.	Thaumaturgical device	123
43.	Iron crucibles	125
44.	Absorption bottles	127
45.	Bottle with a tap	129
46.	Graduated glass vessel	131
47.	Stoppered glass vessel	133
48.	Cucurbits and alembic	135
49.	Bottles and flasks	136
50.	Flasks	137

		Page
51.	Retorts	138
52.	Cucurbit	139
53.	Flasks	140
54.	Bell-shaped vessel	141
55.	Tubes for producing electrostatic charges	141
56.	Instantaneous light machine	148
57.	Mayer's instantaneous light machine, 1811	148

FOREWORD

The Playfair Collection of chemical apparatus was one of the first acquisitions to be made in the field of historic scientific material by the Industrial Museum of Scotland (precursor of the Royal Scottish Museum). The first Director of the Museum, George Wilson, and the Professor of Chemistry at Edinburgh University, Lyon Playfair, showed great foresight in preserving these significant mid-18th to mid-19th century relics of chemistry teaching: in 1858 very little material of this type was thought to be worth collecting.

One of the fundamental tasks of the museum curator is to ensure that the objects in his care are made accessible to others. Published catalogues allow for treatment in depth and this work by Dr Anderson (formerly of the Royal Scottish Museum, now of the Science Museum, London) is the first detailed discussion of this kind. It is hoped that it will be of special interest to those who this year are celebrating the 250th anniversary of the birth of the great Scottish Enlightenment chemist, Joseph Black, with whom part of this Collection is associated.

Norman Tebble DSc
Director

Figure 1 George Wilson (1818-1859), first Director of the Industrial Museum of Scotland. Photograph attributed to John Tunny, ca. 1850
(Reproduced by permission of the Scottish National Portrait Gallery)

PREFACE

At a conference held in 1876 at the South Kensington Museum (from which the Science Museum has evolved), Edward Frankland, speaking with reference to the Special Loan Collection which was then being exhibited, said:

> "The apparatus with which chemists, both ancient and modern, prosecuted their researches was generally of a simple description and often dismantled as soon as the necessary operations were completed, consequently it was far less likely to be preserved than the more expensive and elaborate contrivances of the physicist."[1]

With the advantage of being able to survey an additional century of acquisition and preservation of scientific material by museums and other institutions, Frankland's assessment is still seen to remain true. The number of chemical artefacts collected is small compared with those of physics. This is especially so in the case of pre-19th century apparatus of which very little of known provenance survives. That which does, such as Antoine-Laurent Lavoisier's at Paris and Martinus van Marum's at Haarlem was scarcely that with which their researches were prosecuted. What remains in these collections are specially developed instruments constructed for teaching and demonstration.

The Playfair Collection at the Royal Scottish Museum, Edinburgh, also comes into this category, being part of the working apparatus acquired by the chemistry professors at Edinburgh up to the mid-19th century. Apart from it being much less well-known it differs from the few other collections of its type in that the overall quality of craftsmanship is of a significantly humbler standard. This makes it an even more unlikely survival for the instruments of Lavoisier and van Marum were constructed by the best craftsmen available and they are of outstandingly impressive manufacture. It is likely that the Playfair apparatus was locally made, possibly by artisans who were not professional scientific instrument makers. (Of the 64 items which remain only one is signed by its maker.) This may reflect the paucity of specialist craftsmen in Edinburgh (compared with London and Paris) and the fact that it was purchased by the professors of chemistry themselves from their own incomes.

The Playfair Collection was presented in 1858 by Lyon Playfair to the Industrial Museum of Scotland (now the Royal Scottish Museum). Playfair had just been appointed professor of chemistry at the University of Edinburgh and was making changes to his laboratory when in a store room he discovered apparatus which he thought was of historical significance. He passed it over to his colleague George Wilson (fig 1), the first director of the then new Museum. It was registered as having been the property of Playfair's predecessors in the Edinburgh chair, Joseph Black, Thomas Charles Hope and William Gregory. If this assumption is correct the apparatus must date from the period 1766 or thereabouts (Black's appointment to Edinburgh) to 1858 (Gregory's death). However the grounds for assuming the earlier limit are not clear - chemistry had been taught at the University from about 1713 - and the possibility that part of the apparatus may pre-date Black must be considered.

The charter allowing the establishment of a university at Edinburgh was granted in 1582. It was to be municipally controlled and was permitted to establish all the higher faculties of learning, though it did not immediately exercise this right. Only courses in the liberal arts were offered, taught by non-specialist regents who were responsible for taking a class right through the course to degree level. Some studies related to science and medicine may have been taught incidentally by those regents who were personally interested in these topics (student notes of the 17th century on anatomy are known) though this was probably exceptional. The establishment of the medical school at Edinburgh followed a long period of gestation which started in about 1680. The teaching of chemistry was closely associated with medical studies at the time and a professor of physic and chemistry was appointed in 1713 though it was not until 1726 that the faculty which could award degrees in medicine was set up. The reason for establishing the medical school was partly financial: it was felt that sending Scots to train abroad was economically detrimental to the nation. From the foundation of the faculty chemistry was taught regularly though in the early years the curriculum seems to have been confined to pharmaceutical studies, taught by a consortium which had interests in drug manufacture. This approach was changed in the middle of the 18th century with the appointment of William Cullen who was shortly followed from Glasgow by Joseph Black. For 50 years the teaching they provided was of the highest calibre; students were drawn by their (and their medical colleagues') reputation from all parts of the civilised world. The syllabus which they taught was wide ranging, incorporating what would now be termed industrial chemistry, pharmacy and physics, as well as pure chemistry itself. For that reason the Playfair Collection includes items which would not now be associated with chemistry. Black's successor, Thomas Charles Hope, was in many ways conscientious, though he did not provide sufficient encouragement for practical studies at the very time when the great research schools were being established on the Continent. Edinburgh's prestige declined and ironically Scots students again travelled back to the mainland of Europe to pursue their chemical studies. Hope did not attempt to emulate the European chemists partly because the *raison d'être* for chemistry teaching at Edinburgh was still the medical school. It could be argued that the close association of chemistry and medicine at Edinburgh first promoted chemical studies, but was later responsible for their decline. It was not until 1844 that the title of the chair was changed from that of medicine and chemistry to simply that of chemistry.

This work is divided into two sections. The first attempts to provide a context for the catalogue of the collection which follows. The first chapters are not intended to constitute a history of the teaching of chemistry at Edinburgh or to be an assessment of the lives and achievements of those involved in it. This section deals primarily with the practical aspects: the acquisition of apparatus by the professors, their laboratories and, to a certain extent, what and how they taught. Often evidence for these matters is dishearteningly sparse. The apparatus throughout the period under consideration (1713-1858) was the personal property of the professors of chemistry and thus no official inventories were required by the authorities. The locations of the laboratories can sometimes be traced but details concerning internal fittings and facilities are rarely available. Student lecture notes can provide certain details of courses but they must be treated with some caution. They were often copied (at least, before about 1800) from master sets of notes taken by a professional copyist and were not recorded while the lectures were being delivered. Sketches depicting the demonstrations performed by the professors in front of their students scarcely ever appear. The formal, stilted notes do not resemble *aide memoires* but rather a manuscript version of an unpublished book. As text books became more readily available to students the importance of note taking (or of note copying) declined; thus it seems that no student notes at all survive from the chemistry lectures of William Gregory, professor at Edinburgh from 1844 to 1858. In spite of the difficulties which have been discussed it is possible to associate some items in the Playfair Collection with individual professors with confidence. While the amount of apparatus remaining must form a small fraction of what was acquired over the period, it can help bring to life an exciting phase in the development of chemistry teaching in one of the major centres of learning in Europe.

I must record my gratitude to the many librarians and archivists who have let me consult manuscripts and published works in their care. I wish to thank the following institutions for allowing me to refer to and quote from manuscripts in their possession: Aberdeen University Library; Birmingham Central Libraries; British Library (Reference Division); The Chemical Society; City of Edinburgh District Council Archives; Edinburgh University Library; Glasgow University Archives; Glasgow University Library; The Library Company of Philadelphia; The Trustees of the National Library of Scotland; National Monument Record of Scotland, Royal Commission on Ancient Monuments, Scotland; Otago University Medical School Library; Library of the Royal Botanic Garden, Edinburgh; Royal College of Physicians, Edinburgh; Royal College of Surgeons, Edinburgh; Royal Infirmary of Edinburgh Archives; Royal Scottish Society of Arts; St Andrews University Library; Scottish Record Office; The Trustees of Sir John Soane's Museum; Wedgwood Museum, Barlaston; and the Wellcome Institute for the History of Medicine. Of particular help have been Mrs Pat Eaves-Walton of the Royal Infirmary of Edinburgh Archives, Mrs Elspeth Simpson of Glasgow University Archives and Dr Walter Makey of Edinburgh City Archives. I am most grateful to Sir John Clerk of Penicuik, Miss A Scott-Plummer and Professor D C Simpson for allowing me to consult manuscripts in their personal possession. Lammerburn Press Ltd and the Scottish National Portrait Gallery have allowed me to reproduce illustrations. Dr H J Duncan of the Department of Chemistry, University of Glasgow, kindly performed chemical analyses. Others who have offered helpful comments have been Mr David Bryden of the Whipple Museum of the History of Science, Cambridge, and Dr Frank Greenaway of the Science Museum, London. My greatest debt I owe to Mr Allen D C Simpson of the Royal Scottish Museum who in a multiplicity of ways, from early discussion to final production, has facilitated the publication of this work.

Reference
1. *Conferences Held in Connection with the Special Loan Collection of Scientific Apparatus...Chemistry, Biology, Physical Geography, Geology, Mineralogy and Meteorology* (London 1876) 11.

CHAPTER 1

EARLY CHEMISTRY TEACHING IN EDINBURGH: JAMES CRAWFORD, ANDREW PLUMMER AND WILLIAM CULLEN

The earliest institutional instruction in chemistry in Edinburgh was provided not by the University but by the Incorporation of Surgeon-Apothecaries (the body which later became the Royal College of Surgeons of Edinburgh). This took place in a laboratory built in the closing years of the 17th century. However some form of chemical training was probably available forty years earlier, though this may have been only on a master-apprentice basis for the training of apothecaries. In 1657 the Incorporation of Barber-Surgeons (which had been founded in 1505) sought to assimilate the Edinburgh apothecaries into their body for the control of pharmacy. An Act of the Town Council provided the basis for the association of the professions; the legal document bears sixteen signatures of which four are those of members of the Incorporation including James Borthwick and Thomas Kincaid who styled themselves 'surgeon-apothecary'.[1] Thereafter the body was known as the Incorporation of Surgeon-Apothecaries. In July 1661 the minutes of the Incorporation record that Patrick Cunningham was apprenticed to Borthwick to learn the art of surgery and pharmacy.[2] The latter would have been taught using the facilities provided by a botanic garden (the first in Edinburgh) which had been laid out a few years earlier. In 1656 the Incorporation had purchased the house and yards of Curryhill (adjacent to High School Yards, an area still known by that name). Two years later the sum of £200 Scots was spent on altering the grounds, building walls and planting medicinal herbs and flowers.[3] Though nothing is known of any laboratory facilities which may have existed at this date, the glass works at Leith (the seaport of Edinburgh) was able to supply "clear chemistry and apothecary wares" in 1689 and this would have been a likely source of apparatus used in the production of drugs by the members of the Incorporation.[4]

Formal education was provided by the Incorporation (possibly for the first time) at the beginning of the 18th century. In October 1694 Alexander Monteith (who had been apprenticed to William Borthwick and admitted to the Incorporation in December 1691[5]) petitioned the Town Council for a supply of human corpses for the purpose of teaching anatomy.[6] It is almost certain that Monteith was prompted to this action by the contentious iatrophysicist Archibald Pitcairne, at the time a fellow of the Royal College of Physicians of Edinburgh, but who later was to transfer his allegiance to the Surgeon-Apothecaries.[7] Stimulated into action by Monteith's petition, the Incorporation made a similar request to the Town Council less than a fortnight later. Both petitions were agreed, though in the Incorporation's case a condition was attached: that it "shall before the term of Michaelmass 1697 years build, repair and have in readiness an anatomicall theatre where they shall once a year (a subject offering) have one public anatomicall dissection as much as can be showen upon one body".[8] By October 1696 a new hall incorporating the anatomy theatre was under construction. It is clear that the building incorporated a chemistry laboratory, for the minutes of the Incorporation mention that the chimney of the new laboratory was too small.[9] Monteith had by this time been appointed Deacon. In February 1697 it was reported that the roof would be added shortly and it was decided that the chemistry laboratory would be leased to whomever would offer the highest rent.[10] Two months later Monteith successfully bid the sum of £50 Scots for its use for a period of two years, though it was understood that apprentice apothecaries would be allowed use of its facilities at the time of their examination.[11] Later in the same year he was required to furnish the laboratory with presses [12] and he installed furnaces, paying the sum of £136 12s.0d for bricks, gally-pigs, a 'great hot-pot' and other items, it being agreed that he would be refunded the cost of the furnaces when he vacated the premises.[13] In 1701 Monteith was elected President of the Incorporation and Pitcairne was admitted a fellow after an attempt to reconcile the latter with the College of Physicians following a tumultuous schism.[14] The following year instruction in chemistry and anatomy was offered at the new Hall, the *Edinburgh Gazette* of 23 March 1702 announcing that "the course of chemistrie at the laboratorie in the chyrurgeon apothecaries' Hall, Edinburgh, will begin this year upon Tuesday the fourth day of May". No details concerning this course appear to have survived, and it may have been the only one with which Monteith was associated, for on 8 December 1703 he was paid the sum of £136 Scots by the Incorporation for the installation of his two furnaces, and it may be assumed that he then surrendered his lease.[15]

The basic structure of Monteith's laboratory still survives, though there is now nothing which indicates the use to which it was originally put. Its location in the 1697 Hall is known from a contempoary description, it being "three Rooms in the under story of the house ...... the west end thereof".[16] In addition to his teaching activities there, Monteith may have used the facilities for whisky production. In November 1700 he petitioned the Scottish Parliament that "the art discovered by him to draw spirits from malt equal in goodness to true French Brandie may be declared a manufactory with the same privileges and immunities as are granted to other manufactories".[17] Unfortunately nothing seems to be known of Monteith's more orthodox chemical pursuits. A second course of public dissection was offered in April 1704. This time Robert Eliot, later to be employed as "public dissector of anatomy" by the Surgeon-Apothecaries, took part.[18] Whether another chemistry course was then taught is not clear, though a volume of pressed plant specimens which Eliot presented to the Incorporation in 1702 may indicate an interest in botany and drug preparation.[19]

The College of Edinburgh from which the University evolved was founded in 1582. Classes were first held in Hamilton House, built circa 1554, on a site to the south side of Edinburgh. Other buildings were gradually added (fig. 2). As an institution it was unique among the Scottish Universities (three of which pre-dated its foundation) in that it was supported and controlled by the Town Council — it was commonly referred to as the 'Tounis College'. Municipal patronage continued until 1858.[20] Specialist teaching of medicine and its associated sciences was not introduced until the early part of the 18th century after reforms, including abolition of the 'regenting' system of instruction, were implemented by the Principal, William Carstares.[21] Though three professors of medicine, Robert Sibbald, James Halket and Archibald Pitcairne, had been appointed in 1685 they were unsalaried and probably did not teach in the College. However it is likely that James Sutherland, who was created professor of botany in February 1695, did offer some instruction.[22] He had been leased land adjoining Trinity Hospital in July 1675 by the Town Council for use as a physic garden.[23] In the following year he was awarded a salary and his post was 'joined with to' the other professions taught in the College. Twenty years later he was formally appointed a professor and he was given a supplementary salary for creating a physic garden on land to the east of the College and for teaching two days a week. He was also given a room in the College for keeping books and seeds. In 1702 he also received an appointment from the Surgeon-Apothecaries to instruct apprentices in their garden for a fee of one guinea per student (this was the year in which courses in chemistry and anatomy were first offered by the Incorporation). This arrangement was not to continue for long, however, for a complaint was received from the Deacon that Sutherland was "very much defective in his duty as to the teaching chyrurgeon Apprentices in the Science of bottony" and he resigned his several posts in 1706. Although his enthusiasm had clearly waned by the end of his career, Sutherland can be regarded as the first specialist teacher of science relating to medicine in the University of Edinburgh.

The degree of doctor of medicine was first awarded by the University in 1705, though examination of the candidate was solely by fellows of the Royal College of Physicians of Edinburgh at the request of Carstares; the University was to request the aid of the Royal College in providing examiners until 1726.[24]
Before the medical faculty (which had powers to examine) was set up in that year, the University showed certain indications that it was interested in providing teaching in medicine. In 1713, James Crawford was appointed professor of physic and chemistry. Crawford was a medical graduate of the University of Leyden, though he matriculated only five weeks before he graduated in July 1707.[25] It is not known where he had previously studied; it is possible that he was taught chemistry by Herman Boerhaave at Leyden who at the time was offering private tuition, few pupils attending the official lectures of the Professor of Chemistry, Jacobus Le Mort.[26] He was made a fellow of the Royal College of Physicians of Edinburgh in February 1711 and later in that year was appointed Secretary and Librarian. Carstares wrote to the President of the Royal College in November 1713 asking his opinion of Crawford as a suitable candidate for the proposed post and receiving favourable comment approached the Town Council, who "considering that through the want of professors of physick and Chymistry in this Kingdome the Youth who have applyed themselves to that Study have been necessitate to travel and remain abroad a Considerable time for their Education to the great predudice of the Nation by the necessary charges occasioned thereby" appointed him (though without salary) and allowed him "two convenient Rooms" for teaching.[27] In January 1715 the Council allocated Crawford "an accommodation on and of the ground stories of the Colledge" and allowed the sum of £10 "to put up his stalls and the necessars for his profession".[28] It is thus certain that Crawford did teach; this is confirmed by the survival of two sets of chemistry lecture notes taken by his students.[29] The notes are not entirely similar, the set taken by John Fullerton including a long 'practical' section in which 142 pharmaceutical processes are described, unlike those written by Alexander Monro. That Crawford had a laboratory available is indicated in Monro's notes, where he records Crawford as saying:

> "I shall give a full and clear account of the principles upon which I judge a solid and usefull System of Chymie may be founded, and then faithfully shew you a compleat Series of Chymll. Expernts. in a naturally methodicall order which will divid this Course into two Generall parts, the first containing the Theory, & the other the practice . . .

> The Practicall or Experimentall part shall exhibit as many Experiments as will be sufficient to shew all the most materiall discoveries of Chymy Antient and Modern whither for preparing medecines or for detecting the nature and propties of bodies, but that I may contract that Vast and allmost infinite heap of Experiments, scattered thro' a great number of Authors, into as narrow a compass as possible I shall allwise chuse some one particular bodie out of each class and shew the Experiment upon it and then comprehend the whole operation in some generall rules which may be applyed to every individuall of that class..."

There are few clues to the actual facilities or equipment which Crawford used (none was provided by the Town Council) and the precise location of his laboratory is not clear. In addition to his chemistry lecturing Crawford was required to examine candidates for medical degrees together with the nominated fellows of the Royal College of Physicians, for the first time in November 1718. Examinations now took place in the University Library rather than at the College. Although he was appointed professor of Hebrew in August 1719, Crawford continued to examine until May 1725. Whether he carried on giving his University chemistry lectures is not clear, though on 22 September 1720 a private course of instruction offered by Alexander Monro, Charles Alston and Crawford was advertised in *The Caledonian Mercury*. The course was to begin on 7 November, presumably in the Surgeon-Apothecaries' Hall. Monro was to teach anatomy (he had been appointed professor of anatomy of the Incorporation in January 1720), Alston materia medica (he had held the post of King's Botanist since 1716) and Crawford chemistry. In May 1721 Crawford was appointed to a committee of the Royal College of Physicians to revise the Edinburgh Pharmacopoeia (first published in 1699) for the second edition. In 1725 Crawford was still described as professor of medicine.[30] Thus it appears that he did not abandon his chemical and medical interests following his transfer to the new chair. However from November 1724 four new extra-mural teachers of medicine commenced their activities and it was no longer apt for Crawford to offer instruction.

John Rutherford, John Innes, Andrew St Clair and Andrew Plummer were born within three years of each other (1695–1698) and all studied medicine at Leyden over the period 1718–1722 (though not concurrently). Only Plummer graduated MD at Leyden (in 1722); Rutherford graduated at Rheims in 1719, St Clair at Angers in 1720 and Innes at Padua in 1722. They returned to Edinburgh and were licensed to practice by the Royal College of Physicians in February (St Clair and Plummer) and March (Rutherford and Innes) 1724. It seems certain that their future roles as founder-professors of the Edinburgh medical faculty had been carefully charted by George Drummond (who in 1724 was Second Baillie in the Town Council, and from 1725 to 1727 was Lord Provost[31]) and John Monro, an Edinburgh surgeon-apothecary who had studied at Leyden and who engineered the appointment of his son, Alexander, to a chair of anatomy in the city.[32] The four physicians purchased a house from Joseph Cave, a merchant, for use as a laboratory. Its location was in the north-east corner of the University precinct, bounded to the east by a wall and to the south by the College physic garden (fig. 3).[33] In October 1724 they announced their intention to offer an extra-mural course in the institutions and practice of physic, having teamed up with Alexander Monro and Alston and thus providing a complete course of medicine, which was to commence on 10 November.[34] At the same time they advertised a chemistry course which was to begin in February of the following year:

> "Some Gentlemen of the Royal College of Physicians, who have lately erected an Elaboratory, will begin a Complete Course of Chymistry, with a variety of Experiments on vegetables, Animals and Minerals, according to the Method of the celebrated Herman Boerhaave; where likewise all the Chymical Processes in the New Edinburgh Dispensatory, will be shown as they shall most naturally fall into the Order".

Shortly after this advertisement appeared, the four were admitted fellows of the Royal College, and eight days later, on 11 November 1724, a memorial from them was considered and approved by the Town Council. This requested the use of the College physic garden (which in 1706, following its neglect by Sutherland, had been put under the charge of Charles Preston, and in 1711 was taken over by his brother George Preston). It adjoined the house which they had acquired:

> "These Gentlemen having purchased a House for a Chymical Elaboratory Adjoining to the College Garden formerly let to Mr George Preston And finding That the Garden neglected by Mr Preston had for some years lain in disorder Desired of the Honorable Town Councill That they might be allow'd the use of that Ground for the better Carrying on their design of furnishing the Apothecary Shopes with Chymical Medicines and Instructing the Students of medicine in that part of the Science."[35]

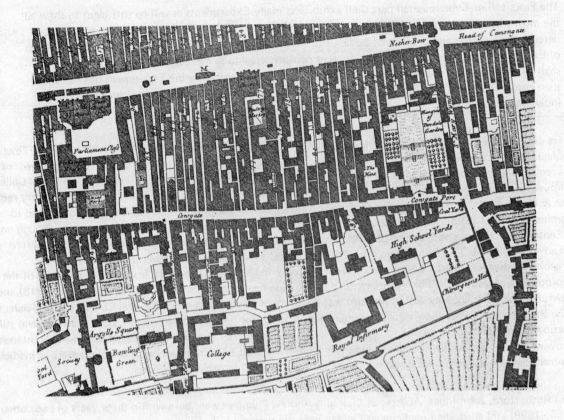

Figure 2 Section from *The Plan of the City and Castle of Edinburgh* by William Edgar, 1742. The High Street runs east-west across the top (north) of this section, while the College (University of Edinburgh), Royal Infirmary and Surgeons' Hall lie just to the north of the City Wall at the bottom.
(Reproduced by permission of Lammerburn Press Ltd, Edinburgh)

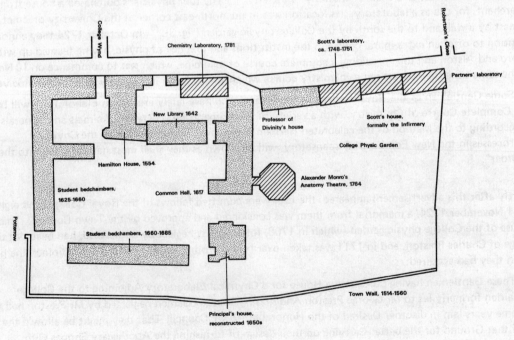

Figure 3 The College buildings prior to 1789, based on the manuscript plan of John Laurie, 1767. Black's chemistry laboratory, built in 1781, has been added.
(With permission of Sir John Clerk of Penicuik)

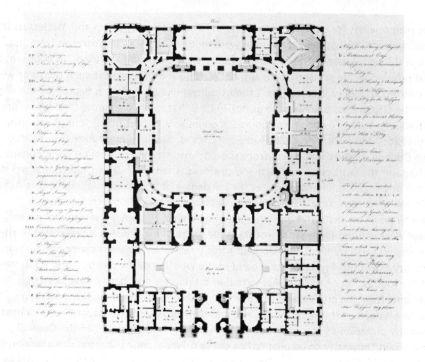

Figure 4 Robert Adam's *Plan of the Principal Story of the New Building for the University of Edinburgh*, engraved by Harding, 1791. The chemistry classroom (K), preparation room (L) and professor of chemistry's house (M) are indicated on the north side (to the right on the plan) between two quadrangles.
(Reproduced by permission of Lammerburn Press Ltd, Edinburgh)

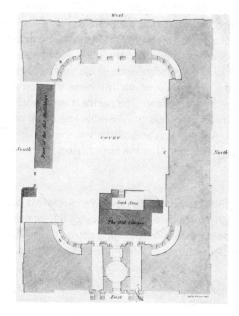

Figure 5 *Plan of the College of Edinburgh as it appears at present*, engraving by W H Lizars, [1823]. William Playfair's simplified adaptation of Adam's building (Figure 4) is not yet complete, part of the Old Common Hall and the student bedchambers still remaining (Figure 3). Hope's chemistry accommodation in the south-west corner (marked B) was occupied from 1823.
(Reproduced by permission of Lammerburn Press Ltd, Edinburgh)

The lease was granted to the memorialists for a period of ten years. "Magistrates, Ministers and Masters of the College" were to be permitted access to the garden, the lessees were required to make an entrance from the laboratory into the garden and the under-part of the gardener's house was to be leased to them, in addition, by the College. While the memorial referred to the philanthropic aim of "nursing and propagating such plants as were necessary for the Improvement of Chymistry in this place", their business-like attitude is reflected by the request "They might have a Grant of the Ground for ten years, so as not to be in hazard of losing the Charges before the Ground cou'd make any suteable return". The promised course of chemistry, taught by all four physicians, commenced on 18 February 1725.[36] In February 1726 Rutherford, Innes, St Clair and Plummer presented a further petition to the Town Council, craving that the patrons should "institute the profession [of medicine] in the Colledge of Edinburgh and appoint the petitioners to teach and profess the same".[37] The Council granted them this request and gave them the power to examine candidates for the degree of doctor of medicine. This Act marks the foundation of the Edinburgh medical faculty. The appointments were for life, but were unsalaried. Rutherford and St Clair were to teach the institutes and practice of medicine, Plummer and Innes chemistry. However only two of the four were granted the right of voting at meetings of the Senatus Academicus in any one year, though this restriction was repealed by the Town Council three years later. Again it seems certain that John Monro and Drummond (by now Lord Provost) had a hand in the wording of the petition and the timing of its submission. The success of the foundation of the medical school at Edinburgh may partly have been due to the favourable financial arrangements from the city's point of view: the four professors of medicine were unsalaried as were Alexander Monro who now taught anatomy in the College and Charles Alston who taught materia medica and botany. The only outlay was the provision of accommodation for Monro — in October 1725 the Council "appropriated a fit place in the said University to be adapted to the said [anatomical] theatre"; this was probably situated at the southern end of the 1617 Great Hall.[38] On the credit side, not only did Scottish students no longer have to pay for their training abroad, but a considerable income was expected from foreigners coming to study in Edinburgh. On 12 October 1726 the Senatus confirmed the appointments made by the Council.[39]

It is clear that teaching students was by no means the only activity with which the professors of medicine were involved and indeed it may have occupied only a small proportion of their energies. This is because they had formed a partnership to prepare and sell drugs on a large scale to apothecaries. The business apppears to have prospered. In a letter book kept by the partners (the letters were initially sent by Innes, later by Plummer) orders for chemicals, drugs and chemical apparatus were recorded from September 1726 to January 1742.[40] All were to suppliers in London.[41] The frequency of the orders, and their size, indicates that the business was a flourishing one. On 15 September 1726, 105 items of glassware were ordered, including six receivers of four gallons capacity (used to collect the products of distillation) and seven mortars of one quart capacity. In July 1728 an order for drugs was supplied at a cost of £28 11s 8¾d. On 19 August 1738 Rutherford ordered 56 pounds of cinnamon. These quantities and costs are not atypical: about 36 orders are included in the volume and it is clear that some additional orders were transmitted verbally to dealers through agents in London. The partners were discriminating in the quality of goods supplied. In July 1729 Plummer complained "there was considerable loss of oil of turpentine; the Nitre fell something short of the weight; the Sal. Ammon was very Dirty and the Tartar was not so good as formerly" and in May 1733 the supplier of glassware was instructed "Let the large glasses be as thick and strong as are made".

The courses of study offered by the newly founded medical school were outlined in the advertisements placed annually in the local press. The first announcements were made in *The Caledonian Mercury* of 8 September and 12 September 1726. After mentioning the anatomical, surgical operations and bandaging course of Alexander Monro, and the materia medica, methodus praescribenda and botany courses of Alston, the advertisement continued:

> "The Institutes of Medicine as digested by the celebrated Herman Boerhaave explained & illustrated by Dr Andrew St Clair & Dr John Rutherfoord. Med. Theor. & Pract. P.P.
> The Practice of Medicine or an Explanation of Dr Boerhaave's Aphorisms de cognoscendis & Curandis Morbis, by Dr John Innes & Dr Andrew Plummer Med. & Chem. P.P.
> A Complete system of Chymistry Theoretical & Experimental according to the same Authors Method demonstrated by the said four Professors.
> These three parts of medicine are taught in the Chymical Elaboratory adjoining to the University".

From 1729 the rider "where all sorts of Chemical Medicines thus publickly prepared are sold to Apothecaries" appears in the advertisement.[42] It has been assumed that only Plummer taught chemistry[43], though the wording of the advertisement implies that all four had a hand. There is evidence in a minute book kept by the partners that in the early years various aspects of chemistry were taught by St Clair, Innes and Plummer. Though the actual content of their contributions cannot be ascertained, a minute of 26 January 1731 reads: "That on the second Thursday of March there be to [sic] chemical Theses given out by Dr St Clair, two Chemico: physiological by Innes, Two Physiological by Rutherford and two Practical by Dr Plummer each of these at eight days interval".[44] A four volume set of lecture notes survives taken, presumably, by a student of Plummer.[45] Headed "Opera Chemica A. Plummer" (though this title may have been added some time after the carefully written notes had been copied out) the work describes 213 pharmaceutical preparations and sets twelve problems; these are all written out in Latin. The notes are completed by two lectures in English, "Of Essential Oils" and "Of Fermentation", the former being subtitled "Abstract of a lecture upon Essential Oils from Dr Andrew Plummer, Professor of Chemistry in the University of Edinburgh". It thus appears from this that Plummer was responsible for dealing with preparative chemistry and that the theoretical part of the Boerhaaveian method promised was taught, if at all, by one of the others; and that Plummer taught mainly in Latin but at times in English. Although not good at delivering lectures due to shyness, his command of his subject was appreciated by his students. John Fothergill who studied at Edinburgh from 1734-36 wrote of him "He knew chemistry well. Laborious, attentive, and exact, had not a native diffidence veil'd his talents as a praelector, he would have been among the foremost in the pupils esteem: such was the gentleness of his nature, such his universal knowledge, that in any disputed point of science, the great Maclaurin appealed to him, as to a living library".[46]

Although the content of Plummer's course appears to have been of very largely pharmaceutical application, Plummer's published work is of more general interest. He presented a paper to the Society for the Improvement of Medical Knowledge (founded in Edinburgh in 1731), 'Experiments on the Medicinal Waters of Moffat', which was published in 1733.[47] The Society lapsed but was revived in 1737 as the Society for Improving Philosophy and Natural Knowledge (or Philosophical Society); in January 1738 Plummer read a paper 'Remarks on Chemical Solutions and Precipitations' and in June 1739 'Experiments on Neutral Salts, compounded of different acid liquors, and alcalIne salts, fixt and volatile'.[48] A rearrangement of teaching responsibilities was occasioned by Innes' death in December 1733 or early in January 1734: the remaining partners agreed in August 1734 that Plummer would teach chemistry, Rutherford the practice of medicine and St Clair the institutes (or theory) of medicine.

The arrangements made by the partners proved lucrative: a double income was provided. Plummer's students paid a fee of three guineas per session to observe him preparing drugs which were then sold to apothecaries. A dividend of forty guineas was paid to the partners in 1731 from income from their students. The flourishing state of the drug business is indicated by the preparation of a new impression of a catalogue in 1732 of which 500 copies were ordered.[49] After Innes' death an account of the partners' stock was taken so that his widow's share could be estimated. The total value at 1 July 1734, including "the Elaboratory House, Utensils, Materials, Preparations and outstanding Debts" stood at £963 4s 8¼d. Of this total, the house itself was worth between £120 and £180[50] which indicates that the stock of drugs and apparatus must have been considerable. In May 1738 the lease of the garden was renewed for a period of twelve years, rent free (the previous lease had expired in 1734, though no action was then taken to renew it).[51]

From about 1739 Plummer hired an assistant, James Scott, a druggist, to prepare the experiments for his teaching.[52] It is not quite clear to what degree Scott was involved in actually demonstrating the processes to Plummer's students. However when the partnership's pharmaceutical activities drew to a close in March 1742 Scott and his partner, Gilbert Laurie junior, rented Plummer and company's laboratory for manufacture of drugs on their own account. This arrangement continued until 1746 when Scott and Laurie built their own laboratory at "the Back of the Cross" close to St Giles.[53] Meanwhile, in July 1744, Scott and Laurie petitioned the Town Council for the lease of the house (described as "ruinous") and an adjacent piece of land. The house was "Bounded on the East by the Elaboratory [ie Plummer's] on the west by the Professor of Divinity's house"[54], the dimensions of the house being 45 feet in length and 20 feet in breadth. This description is sufficient to identify it as the central of the three houses along the north-east boundary of the University precinct and to specify the location of Plummer's laboratory. It is likely that Scott and Laurie wished to use the house as a warehouse for their drugs (Scott had earlier used the Royal Infirmary for this purpose). In June 1748 the partnership between the two broke up and the laboratory and equipment were shared; Scott retained the laboratory and Laurie set up his own in Niddry's Wynd. However at some time between then and March 1751 Scott built himself an entirely new laboratory between his house at the head of Robertson's Close (which he had purchased from the Town Council in April 1747) and the professor of divinity's house (the lights of the latter had earlier been protected by the Council).[55] By an act of 1747 Scott was forbidden entry to the College physic garden adjoining his house[56] and in 1752 he was ordered to build a wall between his house and laboratory, and the garden, to shut off all communication.[57] These two buildings owned by Scott (and shown on a manuscript plan of 1767 drawn by the surveyor John Laurie for the Town Council (fig 3)[58]) are both significant in the development of medical teaching in Edinburgh. The "ruinous house" must be the building leased by the Council in January 1729 to a group of subscribers (whose committee included George Drummond) for use as an infirmary. This was the first voluntary hospital in Scotland; its importance lies in the fact that it was attended by students of the newly founded medical school. In 1736 it was incorporated by charter as the Royal Infirmary of Edinburgh. Shortly afterwards a new, purpose-built hospital was constructed nearby to a design of William Adam and in 1741 the original infirmary house was vacated.[59] Thus the house must have lapsed into its ruinous state between then and 1744 when Scott took over the lease. The adjacent building to the west, built by Scott, may have been used by Plummer's successor Cullen as his classroom for chemical instruction from 1756. In the early years the medical school thus centred on the three houses to the north-east of the College physic garden together with Monro's anatomy theatre in the College proper.

Though the drug business of Plummer and company was no longer operational from 1742, the teaching activities continued. St Clair gave classes in the institutes of medicine until 1747 when Robert Whytt was appointed in his place. Rutherford started to give formal clinical lectures in the new Royal Infirmary building from 1748; he continued these by himself until 1757 after which he was joined by others.[60] He finally resigned his chair in February 1766. It seems that Plummer tried to give up his teaching duties in 1750 and 1751. In May 1750 Rutherford informed the Managers of the Royal Infirmary that the partners were willing to sell the laboratory to them for the sum of £180.[61] This offer was not taken up. In 1751 Plummer (by now described as "a very rich man") tried to transfer his teaching duties to Thomas Elliot (who had graduated MD at Edinburgh in 1746). Before this could be effected Elliot died, in June or July 1751. Plummer therefore continued to give his annual courses until the summer of 1755 when he appears to have suffered a stroke from which he did not recover. In these last years of his teaching his course was not highly regarded.[62] He died in July 1756 by which time his successor, William Cullen, had been appointed joint-professor.

William Cullen was the first teacher of chemistry in Scotland to extend the scope of the subject beyond its application to pharmaceutical preparation; he was interested in approaching practical problems in areas such as agriculture, bleaching and brewing from a chemical standpoint, and this was reflected to some extent in his course. Cullen studied at the University of Glasgow until 1727 by which time he had completed the arts curriculum.[63] He probably attended public lectures in science when he was in London from 1729. Later he served as a ship's surgeon, returning to Scotland in 1732, when he practised as a physician. From 1734 to 1736 he studied medicine at Edinburgh and eventually graduated MD at Glasgow in 1740. Cullen started teaching, initially extramurally, at Glasgow in 1744; two years later he taught a course of the theory and practice of physic within the University. In January 1747 a lectureship in chemistry was established (the first such independent post in the British Isles) partly with money which was saved by postponing the installation of the professor of oriental languages, Alexander Dunlop.[64] Cullen was appointed lecturer, though it was intended that John Carrick (assistant to Robert Hamilton, professor of anatomy) should actually deliver the lectures. Carrick started the course early in 1747 but he was taken ill shortly afterwards and Cullen taught instead.[65] Carrick's indisposition may have been the cause for Joseph Black (who probably first enrolled in the chemistry course in November 1748) being so closely involved with Cullen and his laboratory. In an autobiographical fragment Black wrote:

> "Dr Cullen about this time began also to give lectures on Chemistry which had never before been taught in the University of Glasgow and finding that I might be usefull to him in that Undertaking he employed me as his assistant in the Laboratory".[66]

Unlike the situation which prevailed at Edinburgh, Cullen was provided with finance to set up his laboratory, though it is likely that he provided a certain amount of 'philosophical apparatus' himself (he had brought some to Glasgow from Hamilton in 1744).[67] In January 1747 an estimate of the cost of setting up a chemistry laboratory for teaching purposes was requested by the Faculty; the sum of £52 was proposed. Cullen and Carrick were asked to advise how this should be spent.[68] The money was obtained from the £30 saved by delaying Dunlop's appointment and £22 raised from university funds. In June 1749 Cullen reported that he had spent the initial sum in "building furnaces and fitting up a Laboratory and furnishing the Necessary Vessels for it". Meanwhile in February 1748 Cullen and Carrick were paid £30 for the maintenance of their laboratory and in January 1750 a further £20. Thereafter Cullen was allowed the sum of £20 annually. In September 1749 Cullen was appointed professor of medicine in place of John Johnstoun, the faculty minute recording "It was reported that Dr Cullen designed to continue his Lessons on Chemistry providing the University would give some allowance to support the necessary expense attending the making chemical experiments and processes."[69] The allowance was continued partly because Cullen's income from students' fees was inadequate to cover the costs of his course. In June 1752 it was pointed out at a meeting of the faculty that "Dr Cullen's scholars in Chemistry this Session were very few, and nothing equal to the labour and Expenses he bestows upon his lessons."[70] No inventory of Cullen's apparatus appears to have survived, though a 'List of the Utensiles in the Laboratory' of 1769 may possibly include items acquired by him.[71] There are few clues to what he purchased or from where he ordered his equipment. A faculty minute of June 1749 refers to "Cucurbits, boltheads and a great many other Instruments" which were considered to be Cullen's property as he had purchased them from his own resources, the University allowance being inadequate.[72] The only positive evidence of the provenance of Cullen's apparatus appears to be a letter of December 1747 from his brother-in-law in London, Walter Johnson, in which glassware and chemical texts ordered by Cullen through him are mentioned. Concerning the glass vessels, Johnson apologised that he found "great Difficulty to get any Body to undertake blowing you, as they are seldom or never us'd here by any of the Chymists".[73] He said that he had asked at Apothecaries' Hall as well as enquiring of several chemists whose workmen could not make them, or who would not fix a price until after they were made; he had finally found a man to make them either in green or flint glass, and he forwarded the prices to Cullen. The vessels sketched by Johnson in the letter (presumably copying Cullen's own designs) consist of a tubulated retort (5/-), a double-necked receiver (3/-), a quilled receiver (3/-), a funnel (2/6) and a connecting tube (1/6). Despite Johnson's comment it would not seem that these were of unusual pattern. The books ordered by Cullen at the same time were presumably for his teaching.[74]

An outline of the content of Cullen's lectures at Glasgow can be ascertained from a printed syllabus of his second series of lectures *The Plan of A Course of Chemical Lectures and Experiments to be given in the College of Glasgow MDCCXLVIII during the Session.*[75] In addition the notes from which Cullen lectured have survived and these together with student lecture notes can be used to reconstruct Cullen's course.[76] The introduction included discussion of the history and the use of chemistry. The remainder of the course was then divided into two main categories, the 'general' and the 'particular' doctrines of chemistry. The former included 'PRIMARY CAUSES of the changes of bodies occurring in chemical operations' with subsections on elective attractions, fire, air and ferments, followed by discussion of the operations and instruments of chemistry with comment on solution, distillation and fusion. The latter comprised three major subdivisions, salts, sulphurs, and waters. In the first, acids and alkalis were treated, the second natural products such as oils, resins, etc, soaps and alcohol and the third salt water and mineral waters. Of the application of chemistry to industrial processes, Cullen only specified soap, brewing, the preparation of spiritous liquors and vinegar in the syllabus. Many other manufactures were dealt with in the course of lectures under the appropriate general category, for instance bleaching under alkaline salts. However, the course by no means dwelt solely on such practical affairs, and Cullen gave a balanced account of philosophical chemistry referring critically to concepts such as those developed by Becher and Stahl on salts, Pott on earths, E F Geoffroy on affinity and Martine on fire. In his introductory notes Cullen discussed his approach with reference to the expectations of his students concerning the practical applicability of the course:

> "Some persons may expect very particular Inquiries on the Subject of certain Arts. Examples of the practice of these and Essays made for their Improvement and these parts they would chiefly attend to but I here must inform such persons that we don't pretend to shew the practice of particular Arts in the way of Trade & Business. Neither have we time for doing so nor could the Expence of the Variety of Apparatus necessary be afforded from our Funds. We propose to explain the genl principles of the Arts depending on Chemistry and we shall often also shew the applications to particular Arts in such manner as the practices in the way of Trade & Business may be more easily understood."[77]

In a similar way Cullen criticised the teaching of chemistry when its purpose was simply a vocational training in drug preparation. In his lecture notes he wrote: "Pharmaceutical Courses of Chemistry have not deserved the place they have hitherto held in our Schools that they are not fitted to lead us to a general knowledge of Chemistry that they are not fitted to engage or facilitate an application to this Study and that upon the whole as such has been generally the manner of teaching Chemistry."[78] In referring to "our Schools" it seems highly likely that one of the places Cullen had in mind was Edinburgh. He had studied there from 1734 to 1736 and almost certainly would have attended Plummer's lectures. Evidence of the content of these is consistent with Cullen's comment and clearly Cullen was making a positive effort to provide a much more broadly based course which hitherto had been unavailable in Scotland.

Another characteristic of Cullen's teaching was the unusual emphasis on practical chemistry. Not only did he demonstrate experiments to his students but he encouraged them to conduct their own work. Indeed he was dissatisfied with the response and lack of appreciation of the facilities offered in one year when he concluded his course with the following words:

Figure 6 William Cullen (1710-1790), engraving by John Kay, 1787.

"... every one therefore should in the first place endeavour by easy experiments to acquire some knowledge in this way. The laboratory has been open to you but I am sorry to find that so few of you have frequented it . . . Any of you that have a desire to go farther in the practice may have all the assistances the laboratory affords and you will perhaps find me at more leisure to assist you than I hitherto could. I have no more to add but to thank you."[79]

From the early years of the 1750s Cullen set his eyes on Plummer's Edinburgh chair and was encouraged in this by his sponsor, the improver Henry Home, Lord Kames. Although at one time he had been anxious to dispose of his teaching duties, Plummer now stubbornly clung to his chair and in 1754 Home suggested that Cullen should start to lecture privately in Edinburgh to gain favour with the Royal College of Physicians. Cullen was reticent about accepting this advice and the problem resolved itself with the onset of the illness in the summer of 1755 which prevented Plummer from continuing his course. Among those suggested who might replace Plummer were Joseph Black (then working as a physician in Edinburgh, see chapter two) and Francis Home (a physician who at the time was interested in improving bleaching and agriculture, and who later became professor of materia medica) as well as Cullen himself. Black was immediately available to start lecturing and he obtained Plummer's permission to use the laboratory, starting his course on 29 October.[80] However three days prior to Black's announcement of the course (which because it was incorporated in the advertisement issued by the established professors of medicine bore a certain authority) James Scott offered to step in to "begin a Course of Lectures and Experiments which Dr Plummer observed, on the 12th Day of November, betwixt the Hours one and two Afternoon, at the Elaboratory near the Royal Infirmary".[81] Whether Scott started his course is not clear. On 19 November the Town Council appointed Cullen joint-professor with the incapacitated Plummer without consulting the latter. No arrangement was made to enable Cullen to use Plummer's laboratory and apparatus and indeed the appointment caused resentment among some of the professors as they considered that Plummer should have been consulted. This mismanagement by the Council was sufficient for Plummer to refuse to co-operate with Cullen, for on 22 November Black wrote to Cullen concerning the problem of accommodation, stating that "as the laboratory is entirely his [Plummer's] property, you need not obtain it this winter, but you may probably get Scott's. His apparatus is probably very imperfect for a course of Philosophic Chemistry, but you need not be anxious, provided your course be better than Plummer's which it is impossible for it not to be".[82] This problem was not quickly resolved, for on 29 December Cullen approached the Managers of the Royal Infirmary, their minutes reading: "Dr Cullen having desired that the managers will allow him the use of the vaulted kitchen in the west wing of the Infirmary for giveing a Colledge of Chymistry upon Such terms as they think proper. They authorize the treas to make a Bargain with him".[83]

Cullen started teaching in Edinburgh on 12 January 1756 "in the Laboratory near the Royal Infirmary".[84] It is more likely that this was Scott's rather than Plummer's, for the breach between Cullen and Plummer was not healed until March. The Lord Provost, George Drummond, then intervened in the continuing row between the Council and Senatus concerning the appointment, and wrote to Cullen mentioning the problem of Plummer's laboratory and its possible purchase for the sum of £120, the value when Plummer bought Rutherford's and St Clair's share.[85] The device by which peace was made was for Cullen and Plummer to resign and both to be reappointed on 10 March with the Senatus' approval. Whether Cullen purchased the laboratory is not clear, though there is slight evidence in favour of his continuing to use Scott's premises: Alexander Kincaid, writing 30 years later, stated that Black (Cullen's successor as professor of medicine and chemistry) taught in the house originally used as the infirmary.[86] Though this house never, in fact, appears to have been used for this purpose at least, up to 1756, it was also owned by Scott, being adjacent to his laboratory and possibly connected internally. It is highly likely that Black inherited Cullen's premises. Plummer died in July 1756 and Cullen became sole professor.

Cullen taught a wide range of subjects at Edinburgh. In addition to his university classes in chemistry he also gave private lectures in the subject and, later, on the principles of agriculture and vegetation.[87] From 1757 he delivered clinical lectures at the Royal Infirmary (together with his colleagues Alexander Monro and Robert Whytt). When Charles Alston, professor of medicine and botany, died in November 1760 shortly after the new session had started, Cullen continued delivering his course of materia medica. As Rutherford's health declined in the 1760s it was expected that Cullen would suceed him as professor of the practice of medicine. There was an added incentive for this move in that it would vacate the chair of medicine and chemistry for Joseph Black who was then teaching at Glasgow. However Rutherford was prejudiced against Cullen and favoured John Gregory, professor of physic at King's College, Aberdeen, to succeed him on his retirement. In the event Rutherford's party prevailed and Gregory was appointed in March 1766. By chance, Whytt died a month later. Though Cullen initially decided not to put himself forward as a candidate for the chair of the institutes of medicine because of what he considered unfair treatment by the Town Council, he eventually consented, on being urged by his friends, partly as it afforded the opportunity to bring Black back to Edinburgh.[88]

Cullen was a popular teacher of chemistry at Edinburgh. For his first course only 17 students registered, but this increased to 59 in the next session and as many as 145 attended one of the ten sessions during which he lectured. Some of his pupils attended several courses. James Anderson, editor of the periodical *The Bee* and author of texts on agricultural topics, said of Cullen's chemistry lectures:

> "Dr Cullen was always at pains to examine his students from time to time on those parts of his course that had already been delivered; and wherever he found any one at a loss, he explained it anew, in a clear, familiar manner, suited to the capacity of the student. On these, and other occasions, he frequently desired that whenever any one was at a loss as to any particular, they would apply to him freely for a solution to their doubts and difficulties."[89]

As in the case of Cullen's Glasgow course, a printed syllabus (of 1765) affords information concerning the content of his Edinburgh teaching.[90] The eighteen year period which separates these two publications reveals that Cullen's approach had somewhat changed. Both courses were divided into two parts, but instead of the 1748 'general doctrines' and 'particular doctrines' headings the 1765 course was treated under 'objects of chemistry' and 'operations of chemistry'. The 'objects' were subdivided into saline, inflammable, metallic, earthy, watery and aerial bodies. The 'operations' part occupies a much larger proportion of the syllabus (23 pages compared with eight) and is subdivided into sections dealing with solution, fusion, exhalation [distallation and sublimation] and the application of fire. It would seem from this latter part Cullen dealt in some detail with practical details. This is confirmed by the examination of surviving student lecture notes from this period.[91] He discussed vessels such as matrasses and florentine flasks, and compared alembics with retorts (which he preferred). He made the point that makers of glass apparatus found it difficult to make alembics fit the necks of cucurbits and complained that London manufacturers could not make tubulated apparatus. This last remark was made in notes taken in the early 1760s and shows that the situation had not improved since 1747 when Cullen's brother-in-law wrote to Cullen "The great Difficulty they [glassblowers] say consists in fitting the Openings, and the Stopper to the Tubulated Report which must be fitted by being ground down".[73] Referring to portable laboratory furnaces first described by Francis Hauksbee and Peter Shaw in 1731 Cullen remarked "I know of none yet have been effectual to our wish".[92] It appears from the notes that Cullen did not utilise sophisticated apparatus but was prepared to improvise with what happened to be available. The chair of medicine and chemistry was unsalaried and the professor had to obtain his own apparatus. No inventories of Cullen's own apparatus survive.

Cullen's last chemistry course was given in the 1765-66 session before he transferred to his new chair. He was not distinguished by his contribution to the chemical literature being the author of only one published chemical paper, 'Of the Cold produced by Evaporating Fluids, and of some Other Means of producing Cold'[93], originally read to the Philosophical Society in May 1755. However his application of chemistry to agricultural and industrial problems together with his novel approach to teaching make him an influential figure in the development of the subject. It was said of him "he was the first person in this country [Scotland] who made the chemistry cease to be a chaos".[94]

NOTES AND REFERENCES

1. C H Cresswell *The Royal College of Surgeons of Edinburgh 1505-1905* (Edinburgh 1926) 36, 112.
2. Royal College of Surgeons of Edinburgh, MS Minute Book of the Incorporation of Barber-Surgeons volume 1 (1581-1666) f304, meeting of July 1661.
3. Cresswell *op cit* (1) 48.
4. J Arnold Fleming *Scottish and Jacobite Glass* (Glasgow 1938) 111.
5. Royal College of Surgeons of Edinburgh, MS Minute Book of the Incorporation of Barber-Surgeons volume 2 (1666-1695) f395, meeting of 22 December 1691.
6. Edinburgh District Council Archives, MS Town Council Record volume 35 f84v, meeting of 24 October 1694. The 'Seal of Cause', the original charter granted to the Guild of Barbers and Surgeons of Edinburgh on 1 July 1505, allowed them "anis in the yeir ane condamprit man efter he be deid to make antomell of . . ." (see J D Comrie *History of Scottish Medicine* volume 1 (London 1932) 163) though it seems that the Incorporation did not regularly take advantage of this privilege.
7. Pitcairne wrote to Robert Gray (of London) on 24 October 1694 "I do propose, if it be granted, to make better improvements in anatomy than have been made at Leyden these thirty years, for I think most or all Anatomists have neglected or not known what was most useful for a physician", see British Library, Department of Manuscripts, Sloane Collection MS 3216 f159.
8. Edinburgh District Council Archives, MS Town Council Record volume 35 f88, meeting of 2 November 1694.
9. Royal College of Surgeons of Edinburgh, MS Minute Book of the Incorporation of Barber-Surgeons volume 3 (1695-1708) f48, meeting of 2 October 1696.
10. *Ibid* f57, meeting of 18 February 1697.
11. *Ibid* f62, meeting of 16 April 1697.
12. *Ibid* f87, meeting of 9 November 1697.
13. Cresswell *op cit* (1) 52.
14. R Peel Ritchie *The Early Days of the Royall Colledge of Phisitians, Edinburgh* (Edinburgh 1899) 179; W B Howie 'Sir Archibald Stevenson, His Ancestry, and the Riot in the College of Physicians at Edinburgh' *Medical History* 11 269 (1967).
15. Royal College of Surgeons of Edinburgh, MS bundle 92 (National Register of Archives (Scotland) inventory).
16. *Op cit* (9) f60, meeting of 26 February 1697. The Hall, which has been entirely rebuilt above the ground floor, is situated in Drummond Street. It is now used by the University.
17. T Thomson (ed) *The Acts of the Parliaments of Scotland* volume 10 (1823) 220.
18. R E Wright-St Clair *Doctors Monro* (London 1964) 24; Eliot was appointed to this post in August 1705. Adam Drummond taught with Eliot from 1708. On Eliot's death in 1715 John M'Gill was appointed jointly with Drummond. Eliot was informally referred to as 'Professor of Anatomy'.
19. This volume, inscribed "E dono Roberti Eliot Collegio Edenburg 1702", is now in the library of the Royal Botanic Garden, Edinburgh (MS G 1265).
20. D B Horn 'The Universities (Scotland) Act of 1858' *Univ of Edinburgh J* 19 169 (1959).

21. Regents were teachers who taught all subjects of the arts curriculum and who remained with the same group of students throughout the course.
22. Harold R Fletcher and William H Brown *The Royal Botanic Garden Edinburgh 1670-1970* (Edinburgh 1970) 11.
23. Edinburgh District Council Archives, MS Town Council Record volume 28, f77v, meeting of 7 July 1675.
24. Ritchie *op cit* (14) 280-304.
25. E Ashworth Underwood *Boerhaave's Men at Leyden and After* (Edinburgh 1977) 99.
26. J W van Spronsen 'The Beginning of Chemistry' in T H Lunsingh Scheurleer and G H M Posthumus Meyjes *Leiden University in the Seventeenth Century* (Leiden 1975) 342.
27. Edinburgh District Council Archives, MS Town Council Record volume 41 f263, meeting of 9 December 1713.
28. *Ibid* volume 42, f7, meeting of 13 January 1715.
29. The lecture notes are located as follows: one set, taken by Alexander Monro (later professor of anatomy at Edinburgh) 1715-1717 is at Otago University Medical School Library, Dunedin, New Zealand (MS M164); it comprises 84ff. The other, taken by John Fullarton (and dated June 1713, though this is apparently in error) is at the Wellcome Institute for the History of Medicine Library, London (MS 2451); it comprises 111ff of which the first 43 deal with the 'theoretical' part of chemistry, the remainder the 'practical' part. For the latter, see Underwood *op cit* (25) 203, note 113.
30. *The Physiological Library begun by Mr Steuart and Some of the Students of Natural Philosophy in the University of Edinburg* (Edinburgh 1725) 6.
31. J B Morrell 'The Edinburgh Town Council and its University, 1717-1766' in R G W Anderson and A D C Simpson (eds) *The Early Years of the Edinburgh Medical School* (Edinburgh 1976) 51 and 59, note 20.
32. Underwood *op cit* (25) 102.
33. The house is described as being at the head of Robertson's Close. Closely contemporary maps (for instance, William Edgar's 'The Plan of the City and Castle of Edinburgh' of 1742) show three houses in this location, to the north of the College physic garden. Rutherford and partners' house can be identified as the most easterly of the three by reference to a minute of the Town Council of July 1744, see ref (54).
34. *The Caledonian Mercury* 29 October, 1 November and 5 November 1724.
35. Edinburgh District Council Archives, MS Town Council Record volume 50, f374, meeting of 11 November 1724.
36. *The Caledonian Mercury* 21 January and 26 January 1725.
37. Edinburgh District Council Archives, MS Town Council Record volume 51, f80, meeting of 9 February 1726.
38. Monro's classroom measured 26½ feet long by 18½ wide; a plan of the internal arrangements is in the Edinburgh District Council Archives. See D B Horn 'The Anatomy Classrooms in the Present Old College' *Univ of Edinburgh J 22* 65 (1965-66).
39. Edinburgh University Library MS Dc. 1.4: A Morgan (ed) *Charters, Statutes and Acts of the Town Council and Senatus* (Edinburgh 1937) 224. The MS minute is reproduced in R G W Anderson and A D C Simpson *Edinburgh and Medicine* (Edinburgh 1976) fig 5.
40. 'Book of Letters and Envoys Belonging to the Elaboratory', MS volume in private collection (Miss A Scott-Plummer).
41. The main suppliers of the partners were Joseph Tomlins (and later his son, Joseph) for drugs, and Glisson Maydwell of The Strand for glassware. In 1735 Rutherford commissioned John Fothergill to supply drugs (*ibid*, entry for August 1735).
42. The advertisement appears in manuscript on the back cover of ref (40) in the 1729 style. This implies that it was composed by the partners rather than by Monro or Alston.
43. Underwood *op cit* (25) 117.
44. Edinburgh University Library MS Gen 1959 'Minutes of the Professors of Medicine and Partners of the Chemical Elaboratory in Edinburgh', minute of 26 January 1731.
45. Royal College of Physicians of Edinburgh MS M8/17-20 (the volumes form part of a collection donated to the College in June 1812 by Andrew Duncan *senior*).
46. John Fothergill 'An Essay on the Character of the late Alexander Russell, MD, FRS' in John Elliot *A Complete Collection of the Medical and Philosophical Works of John Fothergill* (London 1781) 643. The somewhat eulogistic assessment of Plummer must be judged in the light of Fothergill's similarly enthusiastic accounts of Monro, Rutherford, Innes, St Clair and Alston, however.
47. Andrew Plummer 'Experiments on the Medicinal Waters of Moffat' *Medical Essays and Observations 1* 82 (1733).
48. Andrew Plummer 'Remarks on chemical Solutions and Precipitations' *Essays and Observations, Physical and Literary 1* 284 (1754); 'Experiments on Neutral Salts, compounded of different acid liquors, and alcaline Salts, fixt and volatile' *ibid* 315.
49. *Op cit* (44) minutes of 29 March and 29 May 1732. No copy of this catalogue appears to have survived.
50. John Thomson *An Account of the Life . . . of William Cullen 1* (Edinburgh and London 1859) 96. The value of the house was estimated by George Drummond in a letter to William Cullen of February 1756, though Drummond implied that the value was less than it might be as the building could only be used as a laboratory. However Rutherford considered the house worth £180, see ref (61).
51. Edinburgh District Council Archives, Town Council Record volume 59, f103, meeting of 10 May 1738.

52. The evidence is contained in an advertisement announcing a course of lectures by Scott, see *The Caledonian Mercury* 16 October, 18 October and 28 October 1755.
53. *The Caledonian Mercury* 18 September and 21 September 1749.
54. Edinburgh District Council Archives, MS Town Council Record volume 64, f301, meeting of 11 July 1744.
55. Scottish Record Office, Particular Register of Sasines (Edinburgh) volume 306, f254 (15 February 1787). This document contains details of the location of Scott's house and laboratory and their dimensions.
56. *Ibid* volume 133, f22 (22 April 1747).
57. Royal Infirmary of Edinburgh Archives, MS Minute Book of the Infirmary volume 3, f139, meeting of 6 November 1752. The Infirmary obtained use of the College physic garden to grow plants from which it made its own drugs.
58. Scottish Record Office MS RHP 9376 'Plan of the Present College of Edinburgh' (reproduced in Anderson and Simpson *op cit* (39) fig 6). This plan shows four houses to the north-east of the College, while Edgar's map of 1742 (ref (33)) shows three. The new building between the most westerly house (that of the professor of divinity) and the two adjacent houses to the east is Scott's new laboratory.
59. P M Eaves Walton 'The Early Years of the Infirmary' in Anderson and Simpson *op cit* (31) 71.
60. Underwood *op cit* (25) 121.
61. Royal Infirmary of Edinburgh Archives MS Minute Book of the Infirmary volume 3, f53, meeting of 7 May 1750.
62. Joseph Black was particularly scathing of Plummer's teaching (see ref (82)) and his attitude contrasts with Fothergill's favourable assessment (ref (46)). However by the time Black could have had first hand experience of Plummer (the session 1752-53) Plummer's enthusiasm may well have waned and Black's other experience, the chemistry teaching of Cullen, would be difficult to match.
63. A full account of Cullen's life and teaching is given in A L Donovan *Philosophical Chemistry in the Scottish Enlightment* (Edinburgh 1975) *passim*.
64. Glasgow University Archives MS GUA 26639. Faculty Minutes f225, meeting of 5 January 1747.
65. Thomson *op cit* (50) *1* 29.
66. Edinburgh University Library MS Dc.2.76^{8*}; Henry Guerlac 'Joseph Black and Fixed Air' *Isis 48* 128 (1957).
67. Donovan *op cit* (63) 49.
68. Glasgow University Archives *op cit* (64) f226, meeting of 28 January 1747.
69. Glasgow University Archives, Faculty Minutes MS 26640, f14, meeting of 12 September 1749.
70. *Ibid* f87, meeting of 26 June 1752.
71. Glasgow University Archives MS 'List of the Utensiles in the Laboratory as delivered to Dr Irvine — July 13th 1769'.
72. *Op cit* (69) f11, meeting of 26 June 1749.
73. Glasgow University Library MS 2255/27 (letter Johnson to Cullen, 8 December 1747).
74. The books ordered are as follows (the titles have been corrected and place of publication and date added; these were not specified by Cullen):
    Johann Joachim Becher *Physica Subterranea* (probably the edition published at Leipzig in 1703 or 1729 as Cullen specified that it should include the commentary by Stahl).
    George Ernst Stahl *Opusculum Chymico-Physico-Medicum* (Halle 1715 or 1740).
    Johann Bohn *Dissertations Chymico-Physicae* (Leipzig 1685 or 1696).
    Johann Heinrich Pott *Excertitationes Chymicae* (Berlin 1738).
    Additionally Johnson wrote that he could obtain a High Dutch grammar and dictionary which Cullen had requested.
75. The syllabus (among the Cullen papers in Glasgow University Library) is incomplete in its printed form but the manuscript from which it was printed partly survives with it though the total may still not be complete. The title of this syllabus differs from that of the pamphlet described by Thomson *op cit* (50) *1* 35 which reads *The Plan of a Course of Chemical Lectures and Experiments, directed chiefly to the improvement of arts and manufactures, to be given in the Laboratory of the College of Glasgow, during the session 1748*. It is unlikely that Thomson would make up this title himself: either there were two syllabuses printed for the same course, or Cullen gave a second course, possibly with a greater emphasis on the applications of chemistry to industrial processes.
76. William P D Wightman 'William Cullen and the Teaching of Chemistry' *Annals of Science 12* 154, 192 (1956); Donovan *op cit* (63) 93.
77. Glasgow University Library, Cullen papers incomplete MS, first line of page reads "Showy manner I might have noticed the".
78. *Ibid*.
79. *Ibid*, MS headed 'In Mixed Arts'.
80. *The Caledonian Mercury* 21 October 1755. The advertisement reads:
    "The Professors of MEDICINE in the University of Edinburgh are to begin their Lectures on Materia Medica, Anatomy, and the Theory and Practice of Physick, on Wednesday the 29th of October. And Doctor Joseph Black, who is to officiate for Doctor Plummer's, begins a Course of Chemistry in Doctor Plummer's Laboratory".
81. *The Caledonian Mercury* 16 October, 18 October and 28 October 1755.
82. Thomson *op cit* (50) *1* 93 (letter Black to Cullen, 22 November 1755).
83. Royal Infirmary of Edinburgh Archives MS Minute Book of the Infirmary volume 3, f225, meeting of 29 December 1755.
84. *Edinburgh Evening Courant* 3 January 1756. The date of Cullen's first lecture is confirmed in the faculty Minutes of Glasgow University (*op cit* (69) f170) for the meeting of 22 March 1756 when it was announced that Cullen had resigned his chair of medicine at Glasgow, having started teaching chemistry at Edinburgh on 12 January 1756.

85. Thomson *op cit* (50) *1* 94 (letter Drummond to Cullen, 3 February 1756).
86. Alexander Kincaid *History of Edinburgh* (Edinburgh 1787) 185.
87. James Anderson 'Cursory Hints and Anecdotes of the late Doctor William Cullen of Edinburgh' *'The Bee, or Literary Intelligencer 1* 53 1791). The lectures were later published as 'Substance of Nine Lectures on Vegetation and Agriculture Delivered to a Private Audience in Year 1768' *in* Board of Agriculture *Additional Appendix to the outlines of the Fifteenth Chapter . . . on the Subject of Manures* (London 1796).
88. Thomson *op cit* (50) *1* 153.
89. Anderson *op cit* (87) 51.
90. [William Cullen] *Chemistry* ([Edinburgh] 1765); syllabus (title page missing?) bound into an untitled MS set of notes taken at Cullen's 1763-64 chemistry lectures, Library Company of Philadelphia MS Yi2.1602.Q.
91. J K Crellin 'William Cullen & Practical Chemistry' *Actes du XIIe Congrès International d'Histoire des Sciences. Paris* **1968** 6 17 (1971).
92. *Op cit* (90) f292.
93. William Cullen 'Of the Cold produced by Evaporating Fluids and of Some Other Means of producing Cold' *Essays and Observations, Physical and Literary* 2 145 (1756).
94. Anderson *op cit* (87) 164.

CHAPTER 2

JOSEPH BLACK

Joseph Black was born near Bordeaux in France in April 1728, the son of an expatriate Ulster wine merchant and a Scottish mother.[1] At the age of twelve he was sent to school in Belfast with two of his brothers.[2] He entered the University of Glasgow in 1744 to commence the arts course (though the first record of his matriculation is two years later). He followed the standard arts syllabus of classical languages and philosophy, the final year of his course being devoted to natural philosophy, this subject being taught by Robert Dick. John Robison, commenting on early influences on Black when introducing the posthumous edition of Black's *Lectures,* mentioned that Dick "was of a character peculiarly suited to Dr Black's taste, having the clearest conception and soundest judgement . . . the most perspicuous and instructive lecturer".[3]

Having completed his arts course in 1748 Black's father urged his son to choose further study which would lead to a profession. He opted for medicine and was taught anatomy by Robert Hamilton and the theory and practice of physic by William Cullen. The latter was to inspire Black to become a chemist and was also to enjoy a close life-long friendship with him. For Cullen's second course of chemistry at Glasgow starting in the autumn of 1748, Black acted as his assistant (see chapter one). In 1752 Black left Glasgow for the University of Edinburgh to complete his medical studies. The reason for this move is not certain, though the Edinburgh medical school was held in greater esteem than that of Glasgow. In Edinburgh Black lived with his cousins Adam Ferguson and James Russell, both of whom were later appointed to the chair of natural philosophy. Robison said of Russell that "no man saw more clearly the great scale of Nature as it is diversified by the powers of mechanism, chemical affinity, and the principles of growth, life, sentiment and intellect"[4] and Dugald Stewart, the philosopher, recalled that Russell's lectures were "comprehensive discussions concerning the objects and the rules of experimental science, with which he so agreeably diversified the particular doctrines of physics."[5] Thus as a student in both Glasgow and Edinburgh Black came under the close influence of teachers whose interests were, for the period, strongly biased towards the practical and experimental aspects of their science.

One of the requirements for the degree of doctor of medicine at Edinburgh (though not at Glasgow) was the submission of a printed thesis. Though Black had intended to get down to this research when he arrived in the autumn of 1752 he delayed doing this for a number of reasons, among which he listed "the shop, infirmary and private patients".[6] His thesis, when he eventually made time for this work, was based on an interest in the chemical properties of limewater which was used medically to dissolve urinary calculi.[7] There is evidence that Black's earliest research, perhaps started at Glasgow, may have been on chalk and lime.[8] However as the medical efficacy of quicklime prepared from different sources was the subject of vigorous dispute between Charles Alston, professor of botany, and Robert Whytt, professor of the theory and practice of medicine, Black judiciously chose to base his research on a substance which resembled the calcereous earths, magnesia alba (basic magnesium carbonate). In the course of his work, he found that on heating magnesia alba a gas was expelled which he suspected had originated in the pearl ashes (potassium carbonate) used in its preparation. By means of a cyclic scheme of quantitative experiments (the first of its kind) Black showed that the original weight of magnesia alba could be recovered by dissolving the product of heating, magnesia usta (magnesium oxide), in sulphuric acid and reconstituting the magnesia alba as a precipitate by the addition of a solution of fixed alkali, suggesting that the origin of the gas, or fixed air, was indeed the alkali. Because the work was being submitted for a medical degree, he felt obliged to include in his thesis a discussion on the effect of magnesia alba on the acidity of the stomach and as a purgative, but he later discarded this relatively trivial part. The work was finally published in June 1754 as *De Humore Acido a Cibis Orto et Magnesia Alba*.[9] It was dedicated to Cullen and Black immediately sent him a dozen copies, enquiring whether he thought it worth publishing in a journal.[10]

From mid-1754 to mid-1755 Black remained at Edinburgh and continued his quantitative research on the reactions of alkalis, though little documentary evidence of this period survives to provide details. In June 1755 he read a revised account of the experiments which formed the second part of his thesis to the Philosophical Society of Edinburgh, together with the results of further investigations into chalk and lime. This paper was published a year later[11] and was his most significant publication. He was able to demonstrate the relationship between quicklime, fixed air and chalk and indicated that fixed air (carbon dioxide) was chemically distinct from atmospheric air. The experiments were simple and used simple apparatus. Black did not describe his apparatus in details but refers to "a flask which was corked close", "a glass retort and receiver", "a Florentine flask" (a long-necked flask of a style similar to those in which oil from Florence was transported) and "a small silver dish over a lamp". The all-important balance on which he performed his weighings (the investigation is sometimes described as the first in which a balance was used at every stage of a planned series of experiments) was not described, though its accuracy can be deduced from the recorded weighings (see catalogue section, object 3). There appears to be no record of where Black performed his experiments, though it seems unlikely that he would have used the laboratory operated by Andrew Plummer which was almost certainly the best available in Edinburgh at the time.

When Cullen moved to Edinburgh in 1756 on his appointment to the chair of medicine and chemistry (see chapter one) Black returned to Glasgow as professor of anatomy and botany and lecturer in chemistry. Here he was immediately burdened with administrative duties[12] though he continued some experiments on alkalis and started a course in chemistry in the following year. He communicated details of an experiment undertaken in 1757 or 1758 (in which he passed carbon dioxide into caustic alkali) to his friend Francis Hutcheson of Trinity College, Dublin, (son of the professor of moral philosophy at Glasgow). These particulars were forwarded to David Macbride, a physician working in Dublin, who published the results and a diagram of the apparatus in his *Experimental Essays*.[13] The diagram is the only illustration of an experimental arrangement devised by Black and published in his lifetime. The apparatus is of extreme simplicity consisting of two bottles joined by a bent glass tube, and a funnel. It is to be assumed that Black inherited the apparatus bought by Cullen from university funds; it is fairly certain that the latter did not transfer much apparatus on his move to Edinburgh as evidenced by the difficulty he found in obtaining the necessary instruments with which to conduct his first course (see chapter one). Even so it seems that Cullen's apparatus was inadequate for Black's teaching as in June 1757 Black was awarded the sum of £40 for additional purchases (in addition to his annual allowance of £20 for teaching).[14] Presumably Black used the laboratory set up for Cullen in 1747, though this later proved to be inadequate and in 1763 a committee recommended that a new laboratory be built and that the old one be converted into a mathematics classroom. In June of that year the committee suggested that the available funds would bear an expense up to £350 and authorised Black to have plans and estimates prepared. However the Principal, William Leechman, and the professor of logic, James Clow, objected to these proposals, saying that the existing laboratory was adequate. Black countered these arguments by complaining that the laboratory was too small, that it was damp and disagreeable, that the floor had never been laid or the walls plastered and that the lecturer had to teach in other rooms while the processes were going on in the laboratory.[15] The Rector, Thomas Miller (who was also Lord Advocate), settled the matter at a meeting held on 2 November 1763 by agreeing that the work should be set in hand, and Black's new laboratory was built.

It was early in Black's career at Glasgow that he first came into contact with James Watt, the engineer. Watt had returned to Glasgow from Greenock after a period of learning the instrument-making trade in London. He operated as mathematical instrument-maker to the University[16], his first recorded transaction with the college being with Black in October 1757 when he provided nine furnace doors.[17] Watt provided Black with a number of miscellaneous items during the following year including a condensing syringe, a lid to a digester, copal moulds, pistons, iron staples and an alarm clock. In November 1758 Black went into partnership with Watt and Alexander Wilson (a typefounder who later became professor of astronomy at Glasgow). Over the next six months Watt charged the partnership the sum of £18 which included 70 days' work though after May 1759 no further mention of the arrangement is found in Watt's account books. Though the partnership seems to have been dissolved, Black still used Watt to construct his apparatus: a ledger kept by Watt from January 1764 lists 24 entries of purchases made by Black, though most of these reflect Watt's activity as a general retailer rather than a purveyor of scientific instruments. The only other known supplier of Black in Glasgow was Ninian Hill, a member of the Faculty of Physicians and Surgeons of Glasgow.[18] Robison reported that Hill had supplied Black with ammonium nitrite prior to 1766.[19]

Black's important work during his second period at Glasgow was on the theory of heat. This interest was probably stimulated by Cullen's work on the lowering of temperature caused by the evaporation of fluids which had been communicated to Black[20] some time before the appearance of his published paper on the subject in *Essays and Observations*.[21] Black had been involved in having an air pump sent to Cullen from Edinburgh in the summer of 1755 for the conducting of his experiments on the lowering of temperature by evaporation.[22] Cullen's work led Black to consider the problem of the heat which was transferred to or subtracted from a body on liquefaction or solidification, but which was not registered by the thermometer. It was Black who introduced the term 'latent heat' and he introduced his ideas early on to his Glasgow classes, possibly during the 1757-58 session. His colleague John Robison suggested that he had developed ideas about different heat capacities of materials leading to the concept of specific heat even earlier, setting out his thoughts in a number of notebooks written before 1757.[23] However this same source suggests that experiments were not performed until 1760. The chronology of his early ideas and experiments is a complex one.[24] From 1764 Black collaborated with a pupil and assistant, William Irvine, in experiments on specific heat[25], work possibly stimulated by earlier work of Herman Boerhaave and George Martine. In the same year Black and Irvine measured the latent heat of steam using crude apparatus and obtained an inaccurate result. Watt performed a similar series of experiments using apparatus which Black regarded superior to his. By the end of 1764 Black is reported to have measured the differences in heat capacities in a variety of substances and communicated the results in his lectures. The work was continued by Irvine after Black moved to Edinburgh in 1766. Black did not publish his theories of heat though urged to do so by several correspondents.[26] A pirated edition of his lectures on heat and chemical apparatus was published in 1770.[27] In a letter to Watt written in February 1783 Black indicated to Watt that he intended to prepare his results for publication in the summer of that year but this promise was not kept.[28]

As in Black's work on alkalis, only the simplest of apparatus was needed for his heat experiments. He referred to such items as cylindrical tinplate vessels, a strong phial and the neck of a broken retort.[29] In Black and Irvine's (and Watt's) measurement of the latent heat of steam a still with a worm condenser fitted into a refrigeratory (a large vessel filled with water) were used. Additionally a balance and thermometers were needed. Black discussed thermometers in his lectures and mentioned them on several occasions in his letters. As well as being supplied by Watt and Wilson (though references to such transactions date from the period after he had left Glasgow in 1766) he probably made thermometers himself. In 1768 Wilson wrote to Black offering to supply thermometers to his students for one guinea and one and a half guineas.[30] Black requested that Wilson send him a thermometer in December 1772 though it was Cullen who required it.[31] In December 1769 Black thanked Watt "for the Thermrs which you blew for me" which he found "most charmingly sized"; however it is clear from the letter that only the ungraduated, unfilled tubes were supplied.[32] Black later asked Watt to "bring when you come, the divided plate for graduating Thermometers".[33] It is possible that Black's "Daimont pencil", bought from Watt in May 1766, was for engraving thermometers.[34] He discussed the blowing of thermometer tubes in his lectures[35] and the notes of one of his students mentions "A Man at Glasgow has made this [the use of a paper scale] unnecessary by engraving on the Tube with a Diamond the proper Degrees of Heat".[36] Another set of notes states that "It is proper to mark the figures on the stem with a diamond but for common use it will be better to have them made like the Glasgow Thermometers".[37]

Black may have been responsible for having a chapter of George Martine's *Essays Medical and Philosophical* of 1740 published in 1772 as *Essays and Observations on the Construction and Graduation of Thermometers*. This reprint was dedicated to Black and was called a second edition. The third and fourth editions of 1780 and 1787 carry similar dedications. The final (1792) edition explains that it is "A New edition with notes and considerable additions, especially the Tables of the Different SCALES OF HEAT exhibited by DR BLACK in his Annual Course of Chemistry". There is also an advertisement: "As this book is recommended by DR BLACK, to the Students attending his class, the Editor has endeavoured to render this addition more useful to them by inserting some notes, and by adding, in the appendix, some tables of the scales of heat which the Doctor usually exhibits and explains in his course".[38] Many of the lecture notes of Black's students contain elaborate folding copies of the comparative scales of temperatures.

There is no item which survives in the Playfair Collection which can be associated with any confidence with Black's work on heat. A thermometer was presented to the Royal Scottish Museum in 1869 by Matthew Forster Heddle, professor of chemistry at St Andrew's University, who stated that it had once belonged to Black though no further details of its provenance was offered (see Appendix 3, object 10). It may be that 'Black's balance' was used in his experiments; this possibility is discussed in the catalogue section (object 3).

By April 1766 arrangements were being made for Black's appointment to the Edinburgh chair of chemistry, Cullen transferring to that of the institutes of medicine (see chapter one). Most Glasgow medical students followed their mentor to Edinburgh. Robison was appointed to the chemistry lectureship at Glasgow. An inventory of the contents of Black's laboratory was made for signature as having been received by his successor; this survives.[39] Before Black departed Robison purchased a quantity of his apparatus for the sum of £12 9s 0d.[40] The last transaction recorded with Black in Watt's waste book is for the supply of a packing box in December 1766, presumably to facilitate the transfer of Black's effects from Glasgow to Edinburgh.[41] Although Glasgow University provided funds for the purchase of chemical apparatus, it is clear that Black owned some of the apparatus he used. In January 1768 he wrote to Watt asking him to send to Edinburgh the copper boiler of a still and some iron pots, and requesting Mr (Ninian?) Hill to sell an oil of vitriol bottle.[42] Black continued to claim his apparatus several years after he had left Glasgow. In February 1770 he asked Watt to send "the light Tongs made for me by Wilson"[43] and perhaps as late as

July 1772 he visited his old laboratory to reclaim "2 lapdals for cucurbits and two separatories" from Irvine (who by now had succeeded Robison as lecturer in chemistry) who noted on an inventory "he said they were part of some Glass Utensils left by him in the Laboratory. I took his receipt for them."[44] Black continued to use the services of Watt as an entrepreneur: in December 1769 he wrote asking Watt to make a pattern from a drawing which he sent him, and then to arrange for John Farey to make a cast of it.[45]

The situation of Black's laboratory and classroom in his early period of teaching at Edinburgh is not altogether clear. On Cullen's resignation he would have almost certainly inherited his facilities, purchasing the apparatus as was the custom. However as already discussed in chapter one, there remains insufficient evidence to decide whether Cullen worked in Plummer's or Scott's laboratory. Alexander Kincaid, writing in 1787, said that Black taught in the house in which the Infirmary was first opened in 1729, that is to say, Scott's house, and though this seems a likely possibility there appear to be no further clues to verify this statement.[46]

In December 1767 the University Senate appointed a committee to discuss with its patrons, Edinburgh Town Council, the complete rebuilding of the College on the same site it already occupied. The instigator of the scheme was William Robertson, the Principal, who outlined his aims in a pamphlet written in 1768:

> "The chief object in view, is, to erect those public buildings which are necessary for an University; a Public Hall, a Library, a Museum, and convenient Teaching Rooms for the several Professors. As there are twenty-one of these, sixteen or eighteen distinct appartments would be required."[47]

At this time the college comprised a medley of buildings (fig 3), the earliest, Hamilton House, dating from 1554, the most recent, Alexander Monro's anatomy theatre, from 1764.[48] The professors, in response to the principal's initiative, submitted plans for their new classrooms. None of these appear to have survived, and the only detail known concerning the proposed chemistry classroom is that it was recommended that a retiring room be provided for Black's personal use.[49] The Town Council endorsed the scheme for rebuilding the college and a subscription list was opened. However offers of financial support fell well short of the estimated cost of £15,000 and the plan was abandoned.

Though this rebuilding plan failed, Black was later provided with facilities in the college. In June 1777 it was reported to the Town Council that a chemistry classroom had been fitted up and that Thomas Heriot, a wright, was to be paid the sum of £45 11s 4d for carrying out the work.[50] This room was situated in a building which also housed natural history and mineralogy specimens and this was to cause difficulties. In 1779 John Walker, minister of Moffat, was appointed professor of natural history. He had accumulated a large collection of specimens which he brought to Edinburgh to use in his teaching. This obviously strained the available accommodation to its limits for in March 1780 Black wrote to the College Committee of the Town Council, saying "He [Walker] and I cannot possibly lodge & arrange our apparatus & carry on our operations in the same Room. I have therefore taken the liberty to look thro' the College for some other place in which I might be accomodated & am humbly of opinion that there is only one which can be fitted up for my use, I mean the Common Hall which is never used at present except twice or thrice a year for public Graduations."[51] Walker himself then sent a report to the Lord Provost of Edinburgh in which he asked if he might move into accommodation adjacent to the library.[52]

It is unlikely that Black obtained permission to use the common hall for in January 1781 he had submitted a "plan of building a Class room" to the college committee.[53] This plan does not appear to have survived either, though certain details can be deduced from the advertisement in *The Caledonian Mercury* of 10 February 1781 which invited tenders for the building work.[54] The location of the building was almost certainly on vacant ground on the north side of Printing House Yard.[55] Black transferred to his new laboratory in the summer of 1781. He wrote to Patrick Wilson (son of Alexander) in April, saying that he was "examining & flitting a vast rubbish of things & experiments from the Old Laboratory" prior to the move.[56]

Although the 1767 attempt to rebuild the college had been unsuccessful, the concept was not abandoned. From 1785 efforts were made to revive the scheme. In 1789 Henry Dundas, member of parliament for Midlothian (and later Home Secretary) tried to raise subscriptions for this venture. Ambitious plans drawn up by Robert Adam were accepted as the chosen design.[57] This time funds were forthcoming. On 16 November 1789 the foundation stone of the east front was laid with great ceremony. Adam's plan was for a double quadrangle, the common side of which was to incorporate a chapel; a number of houses for professors was included. The scheme did not, however, allow for a house for the chemistry professor. This upset Black. He appealed in a letter of December 1789 to the Trustees for Building the New College, arguing that he needed to be constantly at hand to conduct preparations for his class experiments:

> "Dr Black begs leave to suggest that there is no professor whose office stands more in need of this privilege or indulgence than the Professor of Chemistry; he has it is true one hour only of teaching but he must spend several hours every day in his laboratory in preparing for the experiments and operations of the next lecture or in finishing those already begun and as these operations often last ten twelve or twenty four hours, or some of them several days he is under the necessity of looking into it frequently during the day & occasionally must be there early in the morning and late at night nor is this sort of labour confined to the session of the College . . . his office is much more laborious than theirs who have only one hour of lectureing dayly; it is also attended with considerable expence for fewel furnaces Glasses & materials."[58]

It is clear by examining alterations to this letter draft that Black had originally intended to propose to the Trustees that his house should occupy the site reserved for the chapel, though he decided not to include this suggestion in his final letter. However he wrote to Adam surreptitiously at the same time proposing the chapel location for his house and mentioning that he did not approve the innovation of a college chapel which he considered to be "in imitation of the english and foreign Colleges".[59] That Black was successful in his petition is indicated by comparing the manuscript plans drawn by Adam of circa 1789[60] with a published plan of the principal (or ground) floor of the quadrangles of 1791 (fig 4).[61] In both plans Black was allocated a chemistry classroom and an adjacent preparation room, each occupying both the ground and first floor of the north-east corner of the principal quadrangle. In Adam's 1789 plans, the rooms to the west of the chemistry accommodation were allocated to Hebrew and the theory of physic (ground floor) and rhetoric and Belles Lettres, and materia medica classrooms (first floor). By the time the published plan was prepared, these classrooms had been transformed into a house for the professor of chemistry, the chapel remaining in its original position.

The progress of Black's new rooms is recorded in a series of letters from John Paterson, Clerk of Works, to Adam. In March 1790 Paterson indicated that the inclusion of Black's house in the plans was not commonly known: "I shall be careful to say nothing at all about Dr Blacks House to any person himself accepted, and were it not for getting him quicker out of his present Class to make room for his new one and his House I should Say nothing to him more than any other person".[62] By October 1790 the foundation of the new college had been laid westward as far as the end of the chemistry classroom and the external wall had been built up to the first floor.[63] Work then faltered at this part of the building. By March 1791 the foundation had reached Dr Black's house but no more of the wall had been built above ground level.[64] From this stage the whole project began floundering: subscriptions which had been promised were not forthcoming, Adam died in 1792 and a year later attention was diverted by the outbreak of war with France. Building operations had virtually ceased by 1794 when only the east front and the north-west corner of the college had been completed. Where Black conducted his teaching (his 1781 laboratory and classroom had been demolished to make way for his new accommodation) is not certain, though the 1617 common hall and the 1642 new library were possibly the only remaining buildings large enough in which to conduct his class.[65] As far as his personal accommodation was concerned, he lived for the remaining years of his life at 58 Nicolson Street.[66]

Black's scientific work changed in character on his return to Edinburgh. Instead of pursuing the more fundamental problems of philosophical chemistry he took a keen interest in the rapidly developing chemistry-based industries in Scotland. He also became deeply involved in teaching. In a letter to Watt of March 1772 he wrote "I have no chemical News, my attempts in Chemistry at present are chiefly directed to the exhibition of Processes and experiments for my Lectures, which require more time and trouble than one would imagine -"[67], and in February 1786 he explained in a letter to Lord Dundonald that he could not visit his works at Culross because he was in the middle of teaching a course of chemistry and "these are duties we dare not neglect - the students would be dissatisfied & would have a right to complain".[68] Apart from his teaching duties Black found time to carry on a small medical practice[69] and was involved in medical duties at the Royal Infirmary.[70] He was admitted a fellow of the Royal College of Physicians of Edinburgh in May 1767 and was closely involved in the preparation of the 1774, 1783 and 1792 editions of the *Edinburgh Pharmacopoeia*, serving on the revision committee from 1778 to 1780 and again in 1791.[71] Black was President of the College from December 1788 to December 1790 and was appointed Physician to King George III in Scotland. He almost certainly founded the Chemical Society for students in Edinburgh in about 1785[72], he was a member of the Philosophical Society and a fellow of the Royal Society of Edinburgh which succeeded it in 1783.[73]

Black was widely consulted on many chemical topics by those involved in commerce and industry and he undertook experimental work in the investigation of problems which arose from these matters. The extent of his involvement can be ascertained from his correspondence[74], which also provides clues to his unpublished laboratory practices. A topic which gained widespread interest in Scotland in the mid-18th century was that of improving bleaching processes. Black's colleague Francis Home, professor of materia medica, was awarded a premium of £100 in 1756 by the Board of Trustees for Manufactures, Fisheries and Improvements in Scotland for his proposal to substitute sulphuric acid for sour milk in bleaching.[75] Black devised an experimental arrangement to be used in bleaching linen[76] and he contributed an appendix to Home's *Essays on Bleaching* entitled 'An Explanation of the Effect of Lime upon Alkaline Salts; and a Method Pointed Out whereby it May be Used with Safety and Advantage in Bleaching'.[77] He was further involved in improving the Scottish bleaching industry by his analysis of kelp from which potash was obtained; for this, he too was awarded a premium.[78] Black was frequently asked to undertake chemical and mineralogical analyses. As early as 1758, two years after he started teaching at Glasgow, he was asked by Lord Erskine to analyse an ore which he found to contain silver and lead.[79] Black later analysed minerals for the Earl of Hopetoun and his son James who were conducting operations at Lead Hills, Lanarkshire.[80] Other analytical work which he undertook was on water samples, testing the composition of water for the Commissioners of Police of Leith and the Superintendent of Water of Edinburgh, in some senses acting as a public analyst.[81] One of the few scientific investigations published by Black concerned the analysis of water from Icelandic geysers.[82] It was probably for this work that he devised a delicate beam-balance for weighing small quantities of solid matter derived from water samples.[83]

A subject which interested Black for much of his career was the design of furnaces for both laboratory and industrial scale operations. Up to the middle of the 18th century most laboratory furnaces were massive and built into the fabric of the building. Black developed a versatile portable furnace (influenced by that described by Johann Joachim Becher[84]) the basic design of which was to be copied for 150 years. He may have produced them while still a student at Glasgow; a letter of February 1753 to Cullen remarked "I wish to know how you are pleased with the new furnaces. It would give me pleasure to hear they answered entirely your expectation, but if you have any fault to them I want to know it, for my own improvement."[85] They were mentioned by Thomas Reid who attended Black's lectures in Glasgow in 1765[86] and described in lecture notes taken by Thomas Cochrane in Edinburgh in 1768.[87] Black's furnace was fully described in a publication of 1782 by a German who had attended his lectures[88] (though never by Black himself). Letters to Black requesting furnaces imply that he was arranging for their manufacture on a commercial scale[89], and a footnote in the *Edinburgh New Dispensatory* of 1786 mentions that "Those who wish to be provided with Dr Black's furnace, may apply to Mr John Sibbald in College-wynd, Edinburgh. They may be procured, in different sizes, from £1:10s.to £2:10s.price. This gentleman has had the advantage of making these instruments under the immediate inspection of Dr Black."[90] Black also designed a laboratory reverberatory furnace; it appears that these were made at the Carron iron works and cost six guineas.[91] The Board of Trustees asked Black to judge the effectiveness of an air furnace whose construction it was sponsoring[92] and Black later advised his brother, Alexander, on furnaces used in the production of glass.[93] He also took an interest in the iron industry, visiting Carron in 1763[94] and later advising Charles Gascoigne (son-in-law of Samuel Garbett, one of the founders of the works) on the composition of an iron ore.[95] In 1787 Black provided Henry Cort with a testimonial for the products of his technique of producing wrought iron by the reduction of pig-iron with coal in a reverberatory furnace.[96] Black was consulted by Archibald Cochrane, Lord Dundonald, about patenting his method of tar extraction[97] and he provided detailed advice on the economic viability of the scheme.[98] A work on the interrelationship of the tar and iron trades published by Sir John Dalrymple in 1784 was based on Black's opinions and Dalrymple described Black as "The best judge, perhaps in Europe, of such inventions".[99]

In addition to the activities outlined here, Black's correspondence shows that his opinion was sought on further diverse topics such as sugar refining, alkali production, dyeing, brewing, metal corrosion, salt extraction, glass making and vinegar manufacture. As well as advising on industrial matters and performing analyses for those engaged in these pursuits Black also invested in emerging industries. In 1794 he made a loan of £500 to Archibald and William Geddes of the Edinburgh and Leith Glass-house Company[100] and his will indicates that income was owing to him from Robert Graham and other partners of the Culcreach Cotton Company from investments he had made.[101] Of ultimately great significance was the support Black provided for Watt, who required finance for his early steam engine experiments.[102]

Although Black's major contributions to chemistry and natural philosophy were completed by 1766, Black continued to conduct researches of lesser significance during his period of teaching at Edinburgh.[103] These investigations were rarely referred to in his letters and practical matters such as his sources of apparatus were scarcely ever mentioned. As well as needing chemical apparatus for his own laboratory work, Black required such material for his lecture demonstrations (his successor, Thomas Charles Hope, described Black's apparatus as being "very excellent"[104]). There are clues that the instruments used were of a simple nature[105] and Thomas Thomson remarked "He . . . illustrated his lectures by plain and beautiful experiments, the best adapted for the subject under discussion, and just sufficient for the purpose. There was no parade of apparatus, nor brilliant display of showy but useless experiments."[106] No

Figure 7 Joseph Black (1728-1799), engraving by John Kay, 1787

evidence remains that Black acquired his instruments from London or any other major manufacturing centre and it is likely that he employed non-specialist craftsmen. It is probable that Archibald Geddes supplied glassware[107] and it has been mentioned that the whitesmith John Sibbald[108] made Black's furnace but knowing of such a supplier by name is exceptional. Fellow scientists such as Watt[109], Josiah Wedgwood[109a], Alexander Wilson[110] and James Keir[111] sent him instruments from time to time but these were probably not sources on whom Black would have relied for regular supplies. The quality of those items in the Playfair Collection which may be associated with Black indicate that he patronised local men.

Black's reputation in Edinburgh during his professoriate lay largely as a lecturer of exceptional talent. His audience was drawn from a wide area, students coming from Europe, America and Russia.[112] The annual chemistry course ran from mid-November to mid-May. Black lectured five times a week, the course consisting of about 128 lectures. Those attending were by no means only those studying medicine or, indeed, those pursuing a course at the University leading to graduation. Black gave special lectures for lawyers[113], James Boswell among them, who wrote "I attended a course of chemistry by Dr Black between two and three. We were mostly lawyers who agreed to take a course. I did not feel much curiosity for the science."[114] Lectures in science intended for a dilettante public were not uncommon in Scotland towards the end of the 18th century although many itinerant lecturers were more interested in attracting large paying audiences by demonstrating spectacular effects rather than in diffusing scientific knowledge.[115] Sets of notes of Black's lectures written out by students (though usually not at the lectures themselves) were in demand and passed hands at a cost of four or five guineas.[116] Students paid a fee, perhaps to professional note-takers, to borrow notes which they then copied out.[117] Discrepancies between the date which the notes were copied and the text indicate that a master set was sometimes out of date when loaned to a student[118] and care has to be exercised in using notes to date topics which Black was teaching. After Black's death Robison was urged by the executors to prepare an edition of the lectures "to prevent imperfect publications from other hands".[119] Although it was Robison's intention simply to publish Black's notes as he found them, Hope (who at the time had already been lecturing for four years to Black's classes) informed him that "many Scraps were only memorandums"[120] and Robison wrote of the notes that "In places, the train of the subject is interrupted by repetitions from yesterdays lecture, or references to processes going forward in the laboratory - Many of them are on scraps of paper, much altered and interleaved - and the whole are filled with chemical symbols and contractions."[121] Eventually by using Black's notes (written on single octavo leaves and full of erasures, additions and alterations), a set of student notes dated 1773 (which had been in the possession of Black who had made extensive alterations) and by writing a considerable portion himself[122] Robison completed the task. *Lectures on the Elements of Chemistry Delivered in the University of Edinburgh by the Late Joseph Black, M.D. Now Published from his Manuscripts by John Robison LL.D.*, was issued in two volumes in 1803.[123] Considering the problems reported by Robison, too much reliance must not be placed on the text as an accurate transcription of Black's teachings. The three plates showing chemical apparatus are referred to by the editor as "A few figures being added to those expressly alluded to in the Lectures"[124] and provide little information about Black's apparatus.

Black's teaching was carefully prepared and popular. The content of his course tended more towards Cullen's 'philosophical chemistry including its application to arts and manufactures' rather than Plummer's 'chemistry applied to the preparation of drugs', though in 1768 a separate chair of materia medica was established at Edinburgh to deal with teaching of the latter topic. Black's lectures were divided into four main sections, the general effects of heat (expansion, fluidity, inflammation etc), the general effects of mixture, chemical apparatus and the chemical history of bodies. The last and largest section was divided into salts, earths, inflammable substances, metals and waters.[125] The content of the lectures was kept up to date by Black's wide circle of correspondents and his contacts with foreign students among his class, some of whom were already established chemists. It is clear that he read the journals and newly published chemical books and introduced new theories and discoveries into his course. Black devoted several lectures to the developing concepts of chemical affinity and elective attraction to explain chemical reactions. Cullen had devised diagrams to represent attraction between substances and Black adapted these.[126] The question of when he accepted and taught Antoine-Laurent Lavoisier's 'new system of chemistry' which marked the overthrow of the phlogiston theory is problematical. After other European chemists had adopted Lavoisier's doctrines of 1783 wholeheartedly he was reluctant to discard totally the concept of phlogiston to explain combustion, calcination and fermentation.[127] However students at Black's Chemical Society were discussing the new theory enthusiastically in 1785.[128] A letter of October 1790 from Black to Lavoisier marks the approval of the new concepts by Black.[129]

Black's lectures were amply illustrated by demonstrations which he performed on a bench in his lecture theatre. A portrait of 1787 and a caricature of the same year shows him teaching, his apparatus laid out in front of him.[130]

His adeptness at experimental demonstrations was remarked on by a number of writers who attended his lectures, a most vivid account being provided by Henry Brougham who was present at one of Black's last courses:

> "Nothing could be more suited to the occasion; it was perfect philosophical calmness; there was no effort; it was an easy and a graceful conversation. The voice was low, but perfectly distinct and audible through the whole of a large hall crowded in every part with mutely attentive listeners . . .
>
> In one department of his lecture he exceeded any I have ever known, the neatness and unvarying success with which all the manipulations of his experiments were performed. His correct eye and steady hand contributed to the one; his admirable precautions, foreseeing and providing for every emergency, secured the other. I have seen him pour boiling water or boiling acid from a vessel that had no spout into a tube, holding it at such a distance as made the stream's diameter small, and so vertical that not a drop was spilt. While he poured he would mention this adaptation of the height to the diameter as a necessary condition of success. I have seen him mix two substances in a receiver into which a gas, as chlorine, had been introduced, the effect of the combustion being perhaps to produce a compound inflammable in its nascent state, and the mixture being effected by drawing some string or wire working through the receiver's sides in an air-tight socket. The long table on which the different processes had been carried on was as clean at the end of the lecture as it had been before the apparatus was planted on it. Not a drop of liquid, not a grain of dust remained."[131]

Apart from being required to attend lecture courses given by members of the medical faculty, candidates for the degree of MD at Edinburgh had to submit a thesis before they could graduate. Up until now little study has been made into how they chose their subjects or conducted their investigations. Some clearly incorporated the results of experiments undertaken by themselves and would have required laboratory facilities. Black's own work, already discussed, comes into this category as does the research of Daniel Rutherford whose thesis *De Aero Fixo Dicto, aut Mephitico* characterised nitrogen for the first time.[132] On the other hand many students undertook no original work and regarded the production of a thesis as a chore not requiring attention until their *viva voce* examinations were already upon them. Sylas Neville, who graduated in 1775, did not consult Cullen about the subject of his work until 18 June, even though the latest date for submission was 1 August. However he considered that he had applied himself conscientiously to his task, for he wrote in his diary "As I took some pains in the composition of my Thesis, I thought I had the right to take notice of the mean and dishonorable conduct of those who publish the compositions of other men as their own."[133] On consulting Cullen over this point he was told that the Faculty's main cause for complaint was not plagiarism but inadequate command of Latin for which reason two men had been failed that year. It seems that facilities for research could be obtained by those students who sought them. Josiah Wedgwood's third son, Thomas, was clearly engaged in chemical experimentation in Edinburgh (though not for the production of a thesis). In January 1787 he wrote:

> "I have been trying to get the Prussian Acid from the blue. I dissolved it caustic veg.Alk, & Spirit of Wine being added to the solution no precipitate ensued. I thought this was the process I successfully tried at home. Pray set me right in this affair. Bergman is in my Elaboratory . . . I have made good many expts on Prussian Blue which though trifling I shall soon send you an account of."[134]

However where this 'Elaboratory' was situated and how Wedgwood obtained its use is not known; it is not out of the question that Black may have accommodated such students in his own laboratory.

Black always considered his health to be poor and in their letters to one another Black and Watt constantly referred to their own and each other's maladies. In about 1792 John Rotheram (who had graduated MD at Uppsala in about 1775) became assistant to Black, whose health was declining, and delivered a large part of the chemistry course.[135] However he was not to be chosen to succeed to the chair of chemistry and in November 1795 Thomas Charles Hope was appointed conjoint professor with Black. In that year Black still delivered most of the lectures though in 1796 after a flattering introduction by Black, Hope effectively took over the full burden of teaching. Black died in December 1799 in an appropriate and widely described manner.[136] His reputation and influence at his death lay as much in his teaching and advice as in his research, the only paper of significance having been published over 40 years previously.

NOTES AND REFERENCES

1. Much biographical material and a discussion of Black's early career have been brought together in A L Donovan *Philosophical Chemistry in the Scottish Enlightenment* (Edinburgh 1975).

2. Henry Ridell 'The Great Chemist, Joseph Black, his Belfast Friends and Family Connections' *Proc Belfast Nat Hist and Phil Soc 3* 50 (1920).

3. Joseph Black *Lectures on the Elements of Chemistry . . . edited by John Robison 1* (Edinburgh 1803) xx.

4. *Ibid* xxiv.

5. Dugald Stewart *Account of the Life and Writings of Thomas Reid D.D. F.R.S. Edin.* (Edinburgh 1802) 32.

6. John Thomson *An Account of the Life, Letters, and Writings of William Cullen, M.D. 1* (Edinburgh 1859) 574. It is not clear what Black's involvement with a shop was.

7. This work is discussed in some detail by Donovan *op cit* (1) 183 and by Henry Guerlac 'Joseph Black and Fixed Air' *Isis 48* 124, 433 (1957).

8. Black *op cit* (3) *1* xxvi. Robison implies that experiments on fixed air and quicklime were conducted prior to November 1752 (a probable month for Black's move); elsewhere (xxiii) Robison states "it is . . . I think not before the year 1752 that I can date any observations relative to fixed air".

9. Joseph Black *De Humore Acido a Cibis Orto et Magnesia Alba* (Edinburgh 1754); translated by A Crum Brown, see Leonard Dobbin 'Joseph Black's Inaugural Dissertation' *J Chem Ed 12* 225, 268 (1935).

10. Thomson *op cit* (6) *1* 50.

11. Joseph Black 'Experiments upon Magnesia Alba, Quicklime and some other Alcaline Substances' *Essays and Observations, Physical and Literary Read before a Society in Edinburgh 2* 157 (1756).

12. Black *op cit* (3) *2* 87.

13. David Macbride *Experimental Essays* (London 1764). Macbride was examined for the degree of MD at Glasgow by Black; he was awarded the degree on 17 November 1764.

14. Glasgow University Archives, University Minutes GUA 26640,f233 (entry for 17 June 1757): "As Dr Blacks Succes in Teaching Chemistry depends upon furnishing the Laboratory with necessary Implements, which are at present intirely wanting, A sum not exceeding fourty pounds Sterling is allowed for this purpose . . . everything purchased is to remain the property of the University".

15. The arguments for and against the provision of a new laboratory are recorded in the University Minutes, Glasgow University Archives, GUA 26643,f5 (entry for 28 October 1763, Leechman and Clow's case against building a "spacious & ornamental Laboratory") and f8 (entry for 31 October 1763 at which meeting "It appeared to the Majority a Step highly proper & becomeing the present Reputation of this University to further countenance the Study and Teaching of a Science which is one of the most usefull and solid, & which is dayly comeing into greater esteem").

16. There is no satisfactory account of Watt's instrument making activities at Glasgow. Some details are provided by J P Muirhead *The Life of James Watt* (London 1858) and Donovan *op cit* (1) 250 assembles some previously published material. Specific aspects are dealt with by P Swinbank 'James Watt and His Shop' *Glasgow University Gazette 59* 4 (1969).

17. Birmingham Central Libraries, Boulton and Watt Collection M3,MS 'Waste Book James Watt 1757'; the entry for 8 October includes "To the College for turning 9 doors for Furnaces for Dr Blacks Class 2s.0d". See also Swinbank *op cit* (16) 7.

18. Alexander Duncan *Memorials of the Faculty of Physicians and Surgeons of Glasgow* (Glasgow 1896) 256. Hill settled in Glasgow in 1754 and resided in Trongate.

19. Black *op cit* (3) *1* 554.

20. Thomson *op cit* (6) *1* 57 (letter, Black to Cullen, "beginning" of 1755).

21. William Cullen 'Of the Cold Produced by Evaporating Fluids and of some other Means of Producing Cold' *Essays and Observations, Physical and Literary 2* 145 (1756).

22. Thomson *op cit* (6) *1* 579; see also remarks in catalogue section, object 1.

23. Black *op cit* (3) *1* 504.

24. Donovan *op cit* (1) 222; these points are discussed in some detail.

25. Black *op cit* (3) *1* 504.

26. See letters to Black from Martin Wall (15 November 1780), George Buxton (17 November 1788) and Jan Ingenhousz (26 May 1791), Edinburgh University Library MSS Gen 873 volume 1,f103, volume 3,f114 and f205 respectively.

27. *An Enquiry into the General Effects of Heat: with Observations on the Theories of Heat and Mixture* (London 1770).

28. Eric Robison and Douglas McKie *Partners in Science* (London 1970) 123, letter 88 (Black to Watt, 13 February 1783).

29. Black *op cit* (3) *1* 123.

30. Edinburgh University Library MS Gen 873 volume 1 f26 (letter, Alexander Wilson to Black, no date given). Wilson constructed and retailed thermometers in London before returning to Glasgow, see Patrick Wilson 'Biographical Account of Alexander Wilson A.M., Professor of Practical Astronomy in the University of Glasgow' *Trans Royal Soc Edinburgh 10* 287 (1826).

31. Robinson and McKie *op cit* (28) 36, letter 29 (Black to Wilson, 23 December 1772).

32. *Ibid* 22, letter 18 (Black to Watt, 20 December 1769).

33. *Ibid* 39, letter 33 (Black to Watt, 22 May 1773); Black sketched a diagram of this plate, possibly to remind Watt of its appearance.

34. Birmingham Central Libraries, Boulton and Watt Collection M3, MS 'Ledger of Personal Accounts Jany 1764 to May 1769' of James Watt, f5; entry for 30 May 1766.

35. Black *op cit* (3) *1* 326.

36. National Library of Scotland MS 3533,f14 (notes taken by Nathaniel Dimsdale, 1767).

37. British Library Reference Division, Department of Manuscripts, transferred from National Reference Library of Science and Invention, London, MS 40550 (undated, unsigned notes). The evidence of this and the previous reference indicates that a previous estimate of the date of the engraving of thermometer tubes (1800) may be too late, see W E Knowles Middleton *A History of the Thermometer* (Baltimore 1966) 135.

38. George Martine *Essays and Observations on the Construction and Graduation of Thermometers... New Edition* (Edinburgh 1792).

39. Glasgow University Archives MS GUA 43081 'List of Utensils delivered to the Committee appointed to receive the Laboratory from Dr Black', dated Glasgow 25 October 1766. The inventory includes seven distilling furnaces, a Papin digester, a balance with a fifteen inch beam and 413 items of glassware.

40. Glasgow University Archives, University Minutes, MS GUA 26643,f146 (entry of 12 June 1766) "Dr Black represented that there were several Utensils in the Laboratory, which he had purchased with his own Money, and the Meeting judging that these may be still useful in teaching Chemistry, They impower the Committee ... for receiving the Laboratory &c from Dr Black, to purchase those Utensils from the Doctor, if they shall judge it to be for the public Interest". *Ibid* f168 (entry for 23 October 1766) records that the laboratory and college utensils were received from Black, and that he was paid £12 9s 0d for those items of his personal apparatus which he was leaving behind. The inventory of apparatus compiled by the committee (comprising the Principal (William Leechman), Robert Trail, John Anderson, Thomas Hamilton and Alexander Wilson) was signed by Black's successor Robison as being correct and was stored in a charter chest (*op cit* (39)).

41. Watt's 'Ledger of Personal Accounts' *op cit* (34) f5; entry for 3 December 1766.

42. Robinson and McKie *op cit* (28) 15, letter 7 (Black to Watt, 10 January 1768). For further discussions concerning the still boiler see catalogue section, object 23.

43. *Ibid* 23, letter 19 (Black to Watt, 28 February 1770).

44. This note appears on the inventory of apparatus originally compiled for Irvine's receipt on Robison's resignation as lecturer in chemistry in 1769 (Glasgow University Archives MS 'List of the Utensiles in the Laboratory as delivered to Dr Irvine - July 13 1769'). The date of collection of glassware by Black is ambiguous as three different years appear on various dates on the inventory. The receipt has been published (Andrew Kent (ed) *An Eighteenth Century Lectureship in Chemistry* (Glasgow 1950) 143) but the date is misleading (11 May 1773) as it does not appear immediately under the receipt in the original shown in the figure.

45. Robinson and McKie *op cit* (28) 22, letter 18 (Black to Watt, 20 December 1769).

46. Alexander Kincaid *History of Edinburgh* (Edinburgh 1787) 185 ("The house first applied for this purpose [as the Royal Infirmary building] was that formerly used by Dr Black, professor of Chemistry, as the place for delivering his lectures, but since pulled down on account of the building of the South Bridge.") The foundation stone of the South Bridge was laid in 1785, the bridge being completed in 1788. The exact location of this building and the uses to which it was put are discussed in chapter one.

47. [William Robertson] *Memorial Relating to the University of Edinburgh* (Edinburgh 1768) 10; a copy of this rare pamphlet is to be found in the Department of Manuscripts, Edinburgh University Library. For the 1767-68 proposals, see also D B Horn *A Short History of the University of Edinburgh* (Edinburgh 1967) 79.

48. The Town Council hired a surveyor, John Laurie, to make a plan of the college as it was in 1767. This manuscript plan survives and is reproduced in R G W Anderson and A D C Simpson *Edinburgh & Medicine* (Edinburgh 1976) 36, item 138 and fig 6.

49. Edinburgh University Library MS Minutes of the Senatus Academicus, volume 1 (1733-1790), entry for ? March 1768 (" ... it was proposed that the dimensions of the Various Schools or rooms in the intended College for Teaching should be condescended upon ... For the Chemistry Class, see a plan given in by Dr Black ... Each of the Professors sh'd be accomodated with a retiring room adjoining as near as possible to each of their respective Classes.")

50. Edinburgh District Council Archives MS Town Council Record volume 95,f175 (entry for 11 June 1777).

51. Edinburgh District Council Archives MS letter, bundle 16 shelf 36, bay C ('Professor Black to B Steuart' 11 March 1780). The Common Hall was on the ground floor of a building erected in 1617. It can be seen in a lithograph by W Scott Douglas taken from an ink and wash drawing by John Sime (original in the Company of Merchants, Edinburgh) published in *Edinburgh in the Olden Time* (Edinburgh 1880) and titled 'A View of the Inside of the North Part of the College of Edinburgh 1816'.

52. *Ibid* (report from Walker to the Lord Provost and Town Council, 21 March 1780). The room Walker described was 70 feet long by 24 feet wide with a range of ten windows to the south; with reference to Laurie's plan (*op cit* (48)) this can only be situated in the 1642 new library.

53. Edinburgh District Council Archives MS Town Council Record volume 100,f302 (entry for 24 January 1781).

54. *The Caledonian Mercury* 10 February 1781; the building work involved
    "About ten roods ruble work
    250 Feet hewn ribbets, soles and lintels
    540 Ditto hewn corners, skews and cope
    1500 Ditto, ditto, ditto pavement
    100 Ditto, perpendicular vents".

55. The new chemistry classroom can be seen by comparing Laurie's plan of 1767 (*op cit* (48)) with John Ainlie's map of Edinburgh (first state), see William Cowan *The Maps of Edinburgh 1544-1929* second edition, revised by Charles Boog Watson (Edinburgh 1932) 54, map 15^{-2} (this latter can be dated to 1780-1782 because of a dedication to the Lord Provost, David Steuart, who held office in these years). The north side of Printing House Yard (or the old Lower Yard) is shown to be unoccupied on Laurie's plan apart from two small buildings (the Reid Chambers) which were probably in a derelict state. In Ainlie's plan a rectangular building occupies the site, of approximate dimensions 60 x 20 feet. The actual position can be determined because the 1781 classroom itself was demolished in 1790 to make way for Black's new chemistry accommodation to Adam's design. The position of these proposed buildings is indicated on the plan of the new college published in 1791 (ref (61)).

56. Edinburgh University Library MS Gen 872 volume 1,f109 (draft letter Black to Wilson, nd but circa 21 April 1781).

57. In a letter from Adam to Thomas Kennedy of Dunure dated 3 October 1789 (private collection of Lt Col J K MacFarlan, Yeovil: copy in National Monument Record of Scotland, Royal Commission on Ancient Monuments, Scotland) Adam mentions "I have been in Scotland for some Weeks & have made out my Northern excursions. What chieffly brought me down this Autumn was to show a plan I made for a new College & brought with me from London". Arthur T Bolton, in *The Architecture of Robert and James Adam 2* (London 1922) 239, implies that plans by Adam were circulating in 1785 but there seems to be no evidence for this. An alternative scheme dated 15 October 1789 by one Robert Morison is known (National Monuments Record of Scotland, Royal Commission on Ancient Monuments, Scotland, drawing no EDD/220/1).

58. Edinburgh University Library MS Gen 873 volume 3,f174F (draft letter, Black to the Trustees for Building the New College, 30 December 1789).

59. *Ibid* f175F (draft letter, Black to Adam, 30 December 1789). During his career at Glasgow 24 years earlier Black had argued against the provision of a chapel in the College (Glasgow University Archives, University Minutes, MS GUA 26643,f50, entry for 7 June 1765).

60. Sir John Soane Museum, London, drawings by Robert Adam, volume 28, nos 34 and 35 (plans of the principal and first floor of the new college building, undated but circa 1789).

61. 'Plan of the Principal Story of the New Building for the University of Edinburgh' signed 'Robert Adam' and 'Harding Sculpt 1791'.

62. Private collection (Professor D C Simpson) MS letter (Paterson to Adam, 15 March 1790).

63. *Ibid* MS letter (Paterson to Adam, 30 October 1790).

64. *Ibid* MS letter (Paterson to Adam, 2 March 1791).

65. During the last five sessions of Black's professoriate (1794-99) between 221 and 262 students were registered for the chemistry class, see *Evidence, Oral and Documentary, taken by ... the Commissioners ... for Visiting the Universities of Scotland. Volume 1. University of Edinburgh* (London 1837) *Appendix* 130.

66. Wilmot Harrison *Memorable Edinburgh Houses* (Edinburgh and London 1893) 29.

67. Robinson and McKie *op cit* (28) 28, letter 23 (Black to Watt, 22 March 1772).

68. Edinburgh University Library MS Gen 873 volume 4,f29. (letter, Black to Archibald Cochrane, Earl of Dundonald, 17 February 1786).

69. For some details concerning Black's medical work, see William Smellie *Literary and Characteristical Lives* (Edinburgh 1800) 170; Robert Kerr *Memoirs of the Life of William Smellie 1* (Edinburgh 1811) 297; 'Case of Adam Ferguson. Drawn up by Joseph Black M.D. in May 1797' *Medico-Chirurgical Transactions 8* 230 (1816). In a letter from Black to Charles F Greville of 12 October 1790 (British Library, Reference Division, Department of Manuscripts, Add MS 42071,f77) Black comments "during the winter ... the daily business of lectureing and the moderate share of practice which I have in town are as much as I can manage".

70. Black was appointed one of the two ordinary physicians to the Infirmary on 2 October 1775 (Royal Infirmary of Edinburgh Archives, Infirmary Minute Book, volume 4,f349) to replace John Steedman who was in bad health, though Black himself resigned one month later (f350) due to his own ill health. Black served several terms as a Manager, from January 1771 to January 1772 (*ibid* ff233, 256), from January 1774 (f299 possibly continuously to January 1791 (*ibid* volume 6,f73) (though a volume is missing covering the period December 1775 to January 1789), in 1791 (substituting for Andrew Duncan who as President of the Royal College of Physicians of Edinburgh was an *ex-officio* Manager), and again from January 1792 to January 1794 (*ibid* ff121, 165, 188) in his own right.

71. D L Cowen 'The Edinburgh Pharmacopoeia' in R G W Anderson and A D C Simpson (eds) *The Early Years of the Edinburgh Medical School* (Edinburgh 1976) 34.

72. James Kendall 'The first Chemical Society, the First Chemical Journal, and the Chemical Revolution' *Proc Royal Soc Edinburgh 73* 346 (1952).

73. The Philosophical Society and its successor, the Royal Society of Edinburgh, are discussed by Steven Shapin 'Property, Patronage, and the Politics of Science: the Founding of the Royal Society of Edinburgh' *British J Hist Sci 7* 1 (1974).

74. The largest collection of Black's correspondence is to be found in Edinburgh University Library MSS Gen 873-5; an index is available from the National Register of Archives (Scotland), NRA(Scot)/0425. Black's correspondence with Watt is in private hands (D Gibson-Watt) and has been published, see Robinson and McKie *op cit* (28).

75. A and N L Clow *The Chemical Revolution* (London 1952) 182. Cullen had benefited one year earlier for his suggested improvements in bleaching and was (appropriately) awarded three suits of table linen to the value of £21 (*ibid* 177).

76. Edinburgh University Library MS Gen 873 volume 1 f8; see also ff9-11, 15 and 20-21, letters to and from Black of 1763 and 1764 concerning bleaching.

77. Francis Home *Experiments in Bleaching* second edition (Dublin 1771).

78. Andrew Fyfe 'On the Comparative Value of Kelp and Barilla' *Trans Highland Soc 5* 29 (1820); Clow *op cit* (75) 79, 177.

79. National Library of Scotland MS 5077,f211 (letter of Black to Thomas Erskine, 15 November 1758). Black also found cobalt in an ore sent shortly afterwards, see MS 5098,f49 (letter of Black to Erskine, 17 January 1759).

80. Edinburgh University Library MS Gen 873 volume 1,ff28, 29, 31, 34, 40, 41, 43, 45, 51, 57 (letters between Black and John Hope, second Earl of Hopetoun, and James Hope, later third Earl, May 1770 to April 1773). See also National Library of Scotland MS 5099,f133 and Clow *op cit* (75) 364.

81. Edinburgh University Library MS Gen 873 ff36-40F (queries from Leith Police to Black and Robison regarding the quality of Lochend water, and their draft reply, 1782); *ibid* ff182-3 (draft report by Black on the purity of the Water of Leith, July 1782); *ibid* ff186-8 (request to Black from Leith Police asking for report on water from Salsberry [Salisbury] Rock and his draft reply, 14 September 1784); MS Gen 873 volume 4,ff20*-20** (enquiry from Superintendent of Water, Edinburgh to Black, 27 August 1785, and his draft reply).

82. Joseph Black 'An Analysis of the Waters of some Hot Springs in Iceland' *Trans Royal Soc Edinburgh 3* 95 (1794).

83. The balance is described in a letter of Black to James Louis Macie (Edinburgh University Library MS Gen 873 volume 3,f158) and published in *Annals of Philosophy* (New Series) *10* 52 (1825); for further discussion, see catalogue section, object 3, ref (15).

84. Johann Joachim Becher *Tripus Hermeticus Fatidicus* (Frankfurt 1689) 1.

85. Thomson *op cit* (6) *1* 576 (letter, Black to Cullen, 10 February 1753).

86. William Hamilton (ed) *The Works of Thomas Reid DD* sixth edition (Edinburgh 1863) 42 (letter, Reid to David Skene, 20 December 1765). Reid later implied *ibid* 44 that furnaces were not available ready made, only to order (and then with difficulty).

87. Douglas McKie (ed). *Notes from Dr Black's Lectures on Chemistry 1767/8* (Wilmslow, Cheshire, 1966) 17. These notes (from a manuscript in the Andersonian Library, University of Strathclyde) include a sketch of Black's furnace, possibly the earliest illustration of it. A number of later sets of student lecture notes also include similar diagrams, for instance Glasgow University Library MS Ferguson 41,f92 verso "furnace for different operations". A printed class ticket issued for chemistry lectures at Glasgow by Irvine in 1785 includes a representation of Black's furnace, see Kent *op cit* (44) plate XIII.

88. August Christian Reuss *Beschreibung eines Neuen Chemischen Ofens* (Leipzig 1782). The furnace was modified by Frederick Accum, see 'Description of the Portable Furnace constructed by Dr Black and since improved' *Nicholson's Journal 6* 273 (1803), and others. No example of a Black-type portable furnace appears to have survived, though a reproduction constructed from Reuss' instructions is in the Science Museum, London (inventory number 1977-529).

89. Edinburgh University Library MS Gen 873 volume 3,f59 (letter, Franz Schwediaur to Black, 9 November 1787); *ibid* f85 (letter, John Cologan to Black, 5 August 1788) ("He [Dr Luzuriaga] wants besides a Model or rather one of the Furnaces ... to be executed under your direction which I request you will do.")

90. *The Edinburgh New Dispensatory ... by Gentlemen of the Faculty at Edinburgh* (Edinburgh 1786) 50. This is the first edition of the series published at Edinburgh and generally known as *The Edinburgh Dispensatory,* see D L Cowen 'The Edinburgh Dispensatories' *The Papers of the Bibliographical Society of America 45* 85 (1951).

91. See letters from John Cologan to Black (*op cit* (89)), ff85, 94, 97 (5 and 16 August and 5 September 1788).

92. Edinburgh University Library MS Gen 873 volume 1,ff137-145 (letters between Black and Robert Arbuthnot, Secretary to the Board of Trustees, June 1783).

93. Douglas McKie and David Kennedy 'On some Letters of Joseph Black and Others' *Annals of Science 16* 129 (1962).

94. Clow and Clow *op cit* (75) 335.

95. Edinburgh University Library MS Gen 873 volume 3,ff1-4 (letter, Charles Gascoigne to Black, 13 December 1785, and Black's draft reply).

96. Henry Cort *A Brief State of Facts Relative to the New Method of Making Bar Iron with Raw Pit Coal and Grooved Rollers* (1787) 16 ('Extracts from Dr Black's "Remarks on the Experiments made to prove the Strength of Mr Cort's Iron" '). Cort had demonstrated his process to Black and others in 1784 (see Robinson and McKie *op cit* (28) 140, letter 98 (Black to Watt, 28 May 1784) and also Edinburgh University Library MS Gen 873 ff280, 287 (letter, Adam Jellicoe to Black concerning an iron anchor made by Cort, 23 December 1786, and Black's draft reply)).

97. Archibald Clow and Nan L Clow 'Lord Dundonald' *The Economic History Review 12* 47 (1942).

98. *Ibid* 50 (letter, Black to Andrew Stuart (Cochrane's uncle), 25 January 1783).

99. John Dalrymple *Addresses and Proposals on the Subject of Coal, Tar and Iron Branches of Trade* (Edinburgh 1784).

100. Edinburgh University Library MS Gen 874/XI 'Copy of the Minutes taken at the opening of the Repositories of the late Dr Joseph Black'.

101. Scottish Record Office MS 14281/2 (CC 8/8/131/2).

102. Details concerning Black's loans to Watt can be found in Watt's 'Ledger of Personal Accounts' *op cit* (34) and in letters between the two men, see for example Robinson and McKie *op cit* (28) 28, letter 23 (Black to Watt, 22 March 1772) and 40, letter 34 (Black to Watt, 23 February 1774). The loans, made circa 1763-66, exceeded £200.

103. Some of Black's minor researches were published, from his letters, as follows: *Phil Trans Royal Soc 73* 303 (1783) (on the freezing point of mercury); *Die Neuesten Entdeckungen 5* 51 (1782) (Black's (1769) preparation of ethyl nitrate, see also Carl Graebe *Geschichte der Organischen Chemie* (Berlin 1920) 198); *Die Neuesten Entdeckungen 9* 219 (1782) (on the freezing point of water); *Die Neuesten Entdeckungen 10* 140 (1783) (on vitrified sodium ammonium phosphate); *Die Neuesten Entdeckungen 11* 97 (1783) (on reactions of vinegar and other matters); in Tiberius Cavallo *History of Aerostation* (London 1785) (on the hydrogen balloon); in John Sinclair (ed) *The Statistical Account of Scotland 1* 426 (Edinburgh 1791) (on the construction of a kiln for calcining marl); in John Williams *An Account of some Remarkable Ancient Ruins* (Edinburgh 1777) (on vitrified walls in buildings in the Highlands of Scotland). It is probable that Black also had printed *Tables of Preparations of Mercury* and *Tables of Preparations of Antimony*. These sheets of pharmaceutical preparations are found amongst his papers (Edinburgh University Library MS Gen 873 volume 2 f35) and bound into lecture notes of his students (eg Edinburgh University Library MSS Dc.8.155,ff47-8, Dc.8.156,ff2-3 and Gen 48D; and British Library Reference Division MS Add 52495); Robison reprinted them (Black *op cit* (3) *2* 753).

Several of the above publications are simply of letters sent to various correspondents. Robison stated that Black "corresponded occasionally with Seguin, and with Crell, who had been his pupil; but did not encourage much intercourse of this kind, having found that his informations sometimes appeared in print as the investigations of the publishers". (Black *op cit* (3) lxv).

104. *Evidence, Oral and Documentary op cit* (65) *Appendix* 283.

105. See catalogue section for objects 40-63, ref (32). Black appears to have used whatever vessels were to hand in his experiments.

106. Thomas Thomson 'Dr Joseph Black' in *The Edinburgh Encyclopaedia 3* 549 (Edinburgh 1830).

107. Archibald Geddes, at one time a pupil of Black, became his close friend (see Black *op cit* (3) *1* vi,lxx). He was an executor of Black's estate. It is highly probable that he supplied chemical glassware to Black, see catalogue section for objects 40-63, especially refs (18), (26) and (27).

108. John Sibbald was entered on the Edinburgh Burgess Roll as a smith on 19 August 1779. He is recorded as a smith at College Wynd in the *Edinburgh and Leith Post Office Directory* from 1788-90 to 1810-11. He had additional premises referred to, perhaps significantly, as Carron Warehouse, at 42 South Bridge from 1797-98 onwards.

109. Watt supplied Black in 1796 with a portable pneumatic apparatus which he and Thomas Beddoes had developed, see catalogue section, object 26.

109a. Josiah Wedgwood supplied a number of scientists with ceramic ware (free of charge) from the early 1780s. In addition to the well known stoneware mortars he produced retorts, distilling and melting pots, crucibles, filter funnels, syphons, tubes, evaporating pans etc. Joseph Priestley, Matthew Boulton, James Watt and Black himself were amongst those who benefited from Wedgwood's generosity. On 23 September 1785 Black wrote to Wedgwood from Edinburgh:"I wish to have for a particular set of Experiments about a dozen boxes or Vessels which must have the following properties: 1st They must be made of the Black Earthen Ware like the Inkstandishes & glazed on the inside with black Glazeing 2dly The bottoms of these must be square, 2½ Inches each side within, & quite flat 3dly The sides must be upright or at right angles with the bottom & only one inch deep within & as thin as they can easily be made . . . I know that I can apply to a Philosopher who has experience of the value of an apt apparatus on some occasions & who has pleasure in doing kind offices." (Wedgwood Museum, Barlaston (on deposit in Keele University Library) MS 11-30437). These dishes, together with a clay-shrinkage pyrometer of Wedgwood's design, were delivered to Black personally by Wedgwood's eldest son, John, who wrote to his father on 27 October 1785 from Edinburgh (where he was attending Black's lectures) asking that further supplies of the boxes be sent for Black along with samples of clay, two or three nests of evaporating pans, retorts (especially large ones) and pyrometer pieces (*ibid* MS 28-20002). It is not known to what purpose Black was putting the black basaltware boxes.

110. Alexander Wilson, professor of astronomy at Glasgow University, supplied Black with thermometers, see *op cit* (30) and (31). Although no thermometer survives in the Playfair Collection, a mercury-in-glass thermometer with a paper scale (graduated from 0° to 216°F), signed by Alexander Wilson and dated 1782, was acquired in 1975 from the Natural Philosophy Department, University of Edinburgh by the Royal Scottish Museum (registration number 1975.56). Details of its acquisition by the Department are not known.

111. James Keir (MD Edinburgh 1778) author of *A Treatise on the Various Kinds of Permanently Elastic fluids, or Gases* (London 1777) sent Black "an ingenious little apparatus for trying the Effect of diff[eren]t Sub[stan]ces in diminishing Air" prior to 1781, see Robinson and McKie *op cit* (28) 112, letter 81 (Black to Watt, 1 May 1781).

112. Henry Guerlac 'Joseph Black' in Charles Coulston Gillespie (ed) *Dictionary of Scientific Biography 2* (New York 1970) 174; for American medical students at Edinburgh, see Jane Rendall 'The Influence of the Edinburgh Medical School on America in the Eighteenth Century' in Anderson and Simpson *op cit* (71) 95.

113. Basil Cozens-Hardy (ed) *Diary of Sylas Neville 1767-1788* (London, New York, Toronto 1950) 216; the entry for 4 May 1775 reads "This day Dr Black finished his lectures rather unexpectedly, as he had promised to give more pharmacy than usual; but instead of that he gave less, alledging as an excuse that his health required some relaxation. But I cannot help thinking that abridging his proper course after beginning another at an early hour to a set of lawyers &c has not the best appearance".

114. Charles Ryskamp and Frederick A Pottle *Boswell: The Ominous Years 1774-1776* (Melbourne, London, Toronto 1963) 160 (entry for 'Summer Session' 1775). Later in the year Boswell became more receptive: his entry for 18 December (*ibid* 201) records "Dr Black's lecture on gold really laid hold of my attention".

115. John A Cable 'Popular Lectures and Classes on Science in Scotland in the Eighteenth Century' MEd thesis (unpublished), University of Glasgow (June 1971).

116. Black *op cit* (3) *1* xi.

117. John Read *Humour and Humanism in Chemistry* (London 1947) 162; Henry Southeran Limited *Bibliotheca Chemico-Mathmatica. Second Supplement 1* (London 1937) 636, item 10270 (this manuscript is now in the Department of Chemistry, University of St Andrews). Most sets of student lecture notes (of which there are at least 50) which survive are of the 'copied-out' variety. A few seem to have been written while the lectures were being delivered amongst which are those of Thomas Cochrane *op cit* (87) and Thomas Charles Hope (Edinburgh University Library MS Dc.10.9$^{1-8}$).

118. An example of such a discrepancy is found in a set of lecture notes dated 1781 (Edinburgh University Library MS Dc.2.84) in which a student wrote "La Dictionaire Chymie which is a very good book, but I believe not yet translated". This refers to Pierre Joseph Macquer's *Dictionnaire de Chymie* (Paris 1766) (which Black frequently recommended). However the work had been translated in 1771 by James Keir (*A Dictionary of Chemistry* (London 1771)) and it is inconceivable that Black would not know of this especially as Keir was one of his students, graduating in 1778.

119. Robinson and McKie *op cit* (28) 322, letter 213 (Robison to Watt, 18 December 1799).

120. *Ibid* 323.

121. *Ibid* 333, letter 219 (Robison to George Black junior, 20 January 1800).

122. *Ibid* 368, letter 233 (Robison to George Black junior, 9 October 1801); Robison states "my writing was *fully* equal to the doctors papers".

123. In addition to the single British edition, two American and two German editions were published: *Lectures on the Elements of Chemistry* 3 vols (Philadelphia 1806-7) and (Philadelphia 1827); *Verlesungen über die Grundlehren der Chemie aus seiner Handschrift herausgegeben von Johann Robison. Aus dem Englischen übersetzt und mit Anmerkungen versehen von Lorenz von Crell* 4 volumes (Hamburg 1804-5) and (Hamburg 1818).

124. Black *op cit* (3) *2.*

125. J R Partington 'Joseph Black's "Lectures on the Elements of Chemistry"' *Chymia 6* 27 (1960).

126. M P Crosland 'The Use of Diagrams as Chemical 'Equations' in the Lecture Notes of William Cullen and Joseph Black' *Annals of Science 15* 75 (1959).

127. Guerlac *op cit* (112) 180.

128. Kendall *op cit* (72) 355.

129. Letter, Black to Lavoisier, 24 October 1790; translated into French and published in *Annales de Chimie 8* 225 (1791). The original MS survives in the Archives de France, and is published by Kendall *op cit* (128) 354.

130. The portrait (oil on canvas) painted by David Martin in 1787 for the Royal Medical Society (and still in its possession) shows Black holding up a rather substantial 'U' tube, whilst on his lecture bench rest a glass bottle from which emerges a tube, a square bird cage, a candle, an open book and a pair of spectacles; it seems certain that Black is lecturing on respiration, combustion and fixed air. The caricature is an engraving by John Kay dated 1787 published in *A Series of Original Portraits and Caricature Etchings by the late John Kay 1* (Edinburgh 1842) 54. The apparatus depicted is much the same as in Martin's portrait though the effect is less formal, for his equipment is scattered on his bench with odd notes strewn around, a second book and a geological hammer being added. It seems likely that Kay's representation has been adapted from Martin's painting: Black's wig, his quizzical expression, his cuffs, cravat and gown match exactly in both.

131. Henry, Lord Brougham *Lives of Philosophers of the Time of George III* (London and Glasgow 1855) 19.

132. Daniel Rutherford *Dissertatio Inauguralis de Aero Fixo Dicto, aut Mephitico* (Edinburgh 1772). The other, independent, discoverer of nitrogen was Carl Wilhelm Scheele, whose experiments were conducted in 1771 or 1772.

133. Cozens-Hardy *op cit* (113) 223.

134. Wedgwood Museum, Barlaston, (on deposit in Keele University Library), Wd MS Mosley accumulation (letter, Thomas Wedgwood to Alexander Chisholm, Etruria, 11 January 1787).

135. Kerr *op cit* (69) *2* 332. Rotheram edited the fourth, fifth and sixth editions (1794, 1797 and 1801) of the *Edinburgh New Dispensatory,* see Cowen *op cit* (90) 87, 94. He was later appointed professor of natural philosophy at the University of St Andrews.

136. Black died with a glass of milk balanced on his knee. This was reported only days after his death in a letter from Robison to Watt, see Robinson and McKie *op cit* (28) 317, letter 211 (dated 11 December 1799).

CHAPTER 3

THOMAS CHARLES HOPE

Thomas Charles Hope was born in Edinburgh in July 1766, the son of John Hope, professor of botany and materia medica at the University.[1] He was sent to school in Edinburgh and Dumfries.[2] In 1780 he matriculated at Edinburgh University and, as was normal, started by attending the arts classes. In 1782 he was present at John Robison's lectures in natural philosophy and he continued with medical courses. His lecture notes which survive from the period 1782 to 1787 show that he attended the lectures of Joseph Black (chemistry), William Cullen (practice of physic), James Gregory (institutes of medicine), Francis Home (materia medica), Alexander Monro *secundus* (anatomy) and John Walker (natural history) as well as those of Robison.[3] In November 1786, the year before he graduated, his father died and he applied for his chair of botany. Though he had the strong support of Sir Joseph Banks (president of the Royal Society of London) and Sir George Baker (physician to King George III) he was unsuccessful, the chosen candidate being Daniel Rutherford, a pupil of Black and the discoverer of nitrogen. Hope graduated MD in June 1787, the title of his thesis being *De Plantarum Motibus et Vita Complectens;* it is dedicated to Banks and Baker.[4]

In July 1787 William Irvine, lecturer in chemistry and materia medica at the University of Glasgow, died and Hope was appointed in his place in October of that year.[5] He started his teaching career by lecturing in both these subjects during the session 1787-8, but in May 1788 a resolution was carried that "The College having for Many Years looked forward to the Establishment of additional lecturers in the Medical Line; and as the Records bear a Resolution that upon a proper occasion the Lectureship of Materia Medica should be disunited from the Chymistry, the Meeting now think it is proper said Measure should be carried into Execution." Robert Cleghorn was appointed lecturer in materia medica at a salary of £25, Hope being reappointed chemistry lecturer at the enhanced salary of £50.[6] Hope was acquainted with Antoine Laurent Lavoisier's chemical doctrines by the geologist Sir James Hall early in 1787 and possibly he taught the 'new system of chemistry' from this time; in any case Hope visited Paris himself in the Summer of 1788 where he was "well received by M. Lavoisier & his set".[7] Hope made further contacts in the scientific community while travelling through London on his way to France, meeting Sir Joseph Banks, Henry Cavendish, Charles Blagden, and William Herschel, amongst others. In July 1789 Hope's uncle, Alexander Stevenson, who was professor of the practice of medicine at Glasgow, suggested to the faculty that as he had suffered from ill-health for some years his nephew should be appointed his assistant and eventual successor. The faculty agreed to deal with the matter (this involved writing to the Secretary of State of the Home Department as the chair had been established under Crown patronage) but Stevenson himself had pre-empted any such action by having written himself before making the request to the Senate and obtaining approval.[8] The faculty was satisfied, however, by this outcome and Hope became Stevenson's assistant, though continuing to teach chemistry. Hope attained full status as professor of medicine in May 1791 when Stevenson died; Hope then resigned his chemical lectureship.

Hope's major research activity at Glasgow was the characterisation of a mineral, heavy spar, which was found near the village of Strontian in Argyllshire. He found it to contain an 'earth' hitherto unknown to him which he called strontia. Hope stated that he first saw the mineral six years prior to 1793 when it was brought to Edinburgh by a "dealer in fossils".[9] In his lectures he mentioned "When it was first brought to Edinburgh in the years 1786 & 1787, I was convinced it was not the Barytic Earth."[10] He performed a few preliminary experiments on it in Glasgow over the period 1787 to 1791, though he first studied it systematically late in 1791. In March 1792 he described his work to the Literary Society of Edinburgh. Fuller experiments were undertaken and the results were read to the Royal Society of Edinburgh in November 1793, a summary of the paper being published in the *Transactions* published in the following year.[10] The work was published in full in 1798.[11] The question of the precedence of the discovery of strontia has been discussed at some length[12] and Hope may not have been the first to consider it a distinct 'earth'. Adair Crawford, physician to St Thomas's Hospital, London, stated in a lecture of November 1789 that "the substance that has been vended at Strontean for the aerated terra ponderosa... has different properties from that which form the basis of heavy spar"; tests which suggested this was carried out by William Cruikshank.[13] A letter from Crawford to Andrew Duncan, professor of the institutes of medicine at Edinburgh on the same subject was published in 1789.[14] However, Hope does drop a hint that his own work was started earlier than this, when still a medical student. In a footnote to his full paper on strontia, he wrote "The beautiful experiment with the muriate was first mentioned to me in the [year] 1787, by the ingenious gentleman, Mr ASH, who was then studying physic at Edinburgh."[15]

In October 1795 Hope resigned his Glasgow post in favour of the conjoint chair of chemistry at Edinburgh with Black, whose health was deteriorating.[16] In the session 1795-96 Black still delivered the bulk of the lectures. In the following session, after Black had concluded his lectures on heat, Hope was introduced to the class by Black who remarked that he now accepted almost all Lavoisier's doctrines but that his successor had the advantage of having learnt them from Lavoisier at first hand.[17] From then onwards Hope taught the chemistry class (except for twelve lectures 'On the Earths' given by Black). In the Summer of 1797 Hope gave an additional course of lectures.

Figure 8   Thomas Charles Hope (1766-1844), detail from engraving THE CRAFT IN DANGER by John Kay, 1817

On his return to Edinburgh Hope became involved in the institutional and social life of the city. He was made a fellow of the Royal College of Physicians in 1796 and was elected President in 1815, a post which he held for four successive years. Hope probably joined the revision committee of the *Edinburgh Pharmacopoeia* in 1799 and he contributed the chemical part of the ninth and tenth editions of 1803 and 1817.[18] Very shortly after Black's death Hope was elected a manager of the Royal Infirmary.[19] He had been made a non-resident fellow of the Royal Society of Edinburgh in 1788; in 1823 he was elected Vice-President, a post which he held until his death. In 1836 he lectured to the Society on the history of the concept of radiant heat when, in his official capacity, he awarded the Keith Prize to James Forbes.[20]

As Black had done, Hope delivered the six month course of lectures in chemistry from autumn to spring. His course was closely based on that given by Black; Benjamin Silliman, who attended Hope's lectures in 1805 commented "I was forcibly struck with the great resemblance of Dr Hope's lectures, in style, substance, and illustrations, to those of his great master".[21] Records of Hope's style of lecturing vary somewhat. Robert Christison (later professor of medical jurisprudence and materia medica) attended the course in the sessions 1814-15 and 1815-16 and wrote of Hope "His manner and his diction were, indeed, somewhat pompous"[22] However Charles Darwin, who went to Edinburgh in 1825, recorded "The instruction at Edinburgh was altogether by lectures, and these were intolerably dull, with the exception of those on chemistry by Hope."[23] Hope was a conscientious teacher and he kept abreast of advances in his science, communicating these to his pupils. (This was by no means true of all his colleagues: Alexander Monro *tertius* was reputed to use the anatomical notes compiled by his grandfather nearly a century earlier.) The notes from which Hope lectured survive.[24] They are written on octavo sheets slipped into envelopes, one for each lecture; emendations, additions and rejection indicate frequent revision. The presence of printed extracts on early experiments in photography demonstrates that this process continued to the end of Hope's long teaching career. Numbers of students registering for the chemistry class grew to numbers never previously attained. In 1795 when Hope first lectured in Edinburgh, 221 students registered; by 1823 this had more than doubled, to 553. While it is true that it was compulsory for medical students who wished to graduate to attend a six month course in chemistry, reference to numbers gaining the MD degree shows that medical students account for only a proportion of those attending. The anatomy lectures attracted only 240 students in 1823.[25]

The increasing popularity of Hope's course exacerbated the problem of the already inadequate accommodation for chemistry teaching. For nearly thirty years Hope had to lecture in a temporary classroom which had been in use since 1791, the year in which the 1781 purpose-built laboratory had been demolished in preparation for the proposed, but in fact never built, Adam laboratory (see chapter two). Commenting on his arrangements for 1805, Benjamin Silliman noted that Hope "performed all his experiments upon a high table, himself standing on an elevated platform, and surrounded on all sides and behind by his pupils. It was an indifferent room for a laboratory, and the furnace conveniences were very limited".[26] In 1801 Hope had a gallery erected in his classroom to provide additional space for students to attend his lectures.[27] In the following year the Town Council approved a scheme to enlarge the room.[28] Hope's classroom was whitewashed[29] and piped water was supplied to the "large apparatus room" in 1810.[30] By 1811 the number of students registered for chemistry had risen to 473 and an urgent expansion of the classroom was needed. A plan for alterations was remitted to the College Committee in May 1812 by the superintendent of works, Thomas Bonnar.[31] However this was "deemed by him [Hope] inadequate for the accommodation of his Students in respect of seeing the experiments"[32] and an amended plan was agreed without the Council having seen the details (an unusual circumstance) on account of "the advanced state of the season, and the loss of time which would have been incurred in order to carry it into effect." Hope's need for more space must have been very pressing.

Meanwhile, moves were afoot for dealing with the unfinished Adam building. In 1801 a government grant had been obtained for work to prevent deterioration of the sections which had been partly completed. Four years later a public lottery was considered to raise funds for finishing the building. By 1807 it was realised that the cost of completing the College to Adam's design was so great that the raising of the necessary sum was an unrealistic proposition. Between 1809 and 1811 Robert Reid was employed to prepare plans and estimates for completing the building at a realistic cost.[33] For some years it had been appreciated that the extravagant accommodation requested by the professors when Adam had first drawn up plans would have to be scaled down if the building were to be completed. At the request of the Senatus, Andrew Duncan *junior* (professor of medical jurisprudence) circulated a printed letter, *Queries applicable to Professors Class-Rooms and Private Rooms,* in July 1815.[34] This was a request for information about what accommodation was felt to be necessary. The replies were published.[35] Considering the very small number of purpose-built chemistry laboratories existing at the time and the low standard of accommodation which was generally available, Hope's suggestions are of considerable interest. He asked for seven rooms, a classroom, laboratory, apparatus room, a "smaller room for more delicate instruments", museum (for chemical preparations, mineralogical specimens, etc), sitting room (for his private use), and "accommodation for the coarser parts of apparatus". Hope insisted that his classroom and laboratory be well lit and ventilated. He specified that his classroom should provide seating for 500 students ("with ample space left for more seats in case of future demand") and that the lecturing area should be railed off, should have a stone floor, a bench 16 feet long with a water supply laid on, and a hood over the furnace to carry off noxious vapours.[36]

In 1816 it was agreed to exclude houses for professors in the new specification and to reduce Adam's double quadrangle to a single one. Various architects submitted ideas with these requirements in mind. The chosen scheme was that submitted by a young and relatively inexperienced architect, William Playfair.[37] He drew up detailed plans of the chemistry classroom and laboratories between 3 and 10 September 1817 and sited them in the south-west corner of his quadrangle. The accommodation comprised, on the ground floor the lower laboratory and lumber room, on the first floor the upper laboratory, the large apparatus room, the small apparatus room and the professor's retiring room, on the first and second floor the classroom and on the second floor the mineral room.[38] The rooms appear to have been ready for use by the start of the 1823-24 session (fig 5). The suite must have formed what must have been the best teaching facilities for chemistry in the British Isles and also must have compared favourably with the best laboratories anywhere at the time. The classroom was the largest in the College. It could hold 530 students and for many years it served for the annual medical graduation ceremony. Hope must have found his new rooms highly satisfactory; he made no further requests to the College Committee for additions or alterations up to the time he resigned his chair in 1843.

Like his predecessor, Hope was renowned for his lecture demonstrations. Christison referred to Hope's "unexampled splendour and success in experimental demonstration"[39] while Silliman stated that "the experiments were prepared on a liberal scale. They were apposite and beautiful, and so neatly and skillfully performed, that rarely was even a drop spilled on the table."[40] Hope claimed that his daily lecture required several hours preparation and reference to his lecture notes and instructions for his assistant give no reason to doubt this statement. The notebooks giving details of apparatus and preparation for his daily experiments indicate the large number of experiments undertaken and the careful planning required.[41] In the summer months Hope commissioned new apparatus and got broken instruments repaired.[42] It is clear that he lay considerable importance on the lecture demonstration as a means of teaching chemistry, it being said of him that he "devoted his whole professorial life to perfecting his admirable class experiments."[43]

In contrast to his enthusiasm in performing experiments in front of his students Hope discouraged, or at least made little provision for, experimental work by them.[44] Christison wrote that he did not "encourage experimental inquiry among his students. His laboratory was open to no one but his class assistant, Dr Fyfe . . . There was at that time [1814] no opportunity for students to learn chemistry practically, either at the great chemical school of Edinburgh, or, indeed, anywhere else in the United Kingdom".[45] This deficiency was partly filled by extra-mural classes offered by private lecturers though sometimes the students themselves made co-operative arrangements. Christison and ten others formed a chemical society which met once a week when they "performed and demonstrated such of Dr Hope's experiments as were within our means".[46] Among the earlier extra-mural lecturers in practical chemistry were Thomas Thomson (later professor of chemistry at Glasgow University) who taught from 1807 and Andrew Fyfe (later mediciner at King's College, Aberdeen) from 1809. By 1822 Fyfe was offering five practical courses in the summer, all of which were well attended. Hope's initial unwillingness to offer practical courses may have been due to his inadequate facilities. The new Playfair laboratory changed this situation and in April 1823 Hope wrote to the Senatus in the following terms:

> "It has long been my wish to improve the means of chemical instruction in the University by affording to the students opportunities of becoming familiar with the Experiments & operations of Chemistry. A Laboratory was provided for this purpose in the New Buildings, and it is at length duly fitted up. I have the satisfaction of mentioning that my assistant Dr John Anderson is now ready to commence the necessary instruction in practical Chemistry & Pharmacy. I trust this measure will meet with the approbation of the Senatus Academicus . . ."[47]

Hope's request was agreed and he reported to the Senatus in November 1823 that his laboratory would be open from 1.00 to 2.00 pm and from 3.00 to 4.00pm, and also for one hour in the evening. However, Anderson was not allowed to use any University apparatus or materials (of which there would be little available anyway), having to supply these himself. Effectively (as suggested by Anderson's successor, Reid) the assistants were acting as private lecturers in the University laboratory, paying the expenses but receiving only a proportion of the remuneration.[48] These arrangements continued until 1828 and were not popular with the chemistry students, only a tiny fraction of whom paid the additional fee to attend the practical class. In 1828 the situation changed when David Boswell Reid was appointed Hope's assistant.[49] He had taught practical chemistry independently for two years in a laboratory described as "a small and inconvenient room in High School Yards." Reid's first intra-mural course was conducted in a room rented from the University. His ability and enthusiasm resulted in his lectures being well attended. However in 1829 the Royal College of Surgeons of Edinburgh decided that it would be necessary in future for candidates who wished to become licentiates to attend a three month course in practical chemistry.[50] This was a blow to Reid, whom the College did not consider qualified to teach (he had not gained an MD degree), but Hope was able to rectify the situation by means of a subterfuge. His name was inscribed on the class ticket with Reid's and he delivered the opening lecture; the rest was left to Reid. The class was given in Hope's preparation room and was limited to 25 students at a time.[51] Hope spent £153 on new apparatus for this venture but received one-third of the income from the class after expenses had been paid. The first three years of this arrangement produced an income of £1,819 13s. 0d. of which £557 3s. 9d. was considered to be expenses.[52] Reid claimed that he provided a further £300 from his personal income with which he purchased portable apparatus additional to the 'fixtures' which Hope had paid for. Reid soom became dissatisfied with the arrangements — he had gained his MD degree in 1830 which qualified him to give chemistry courses (which would satisfy the Surgeons) independently of Hope — and in 1833 he petitioned the Town Council of Edinburgh to establish a chair of practical chemistry in the University. Inevitably, because it would erode his status and income, Hope opposed the idea. A war of pamphlets followed and Reid's suggestion was narrowly defeated by the Town Council.[53] Reid then resigned and returned to private lecturing, causing further consternation by reducing his fees and undercutting Hope's course and those offered by other private teachers (see chapter four). In 1834 Reid proposed that a university lectureship in practical chemistry, independent of the chemistry chair, be set up. This was again opposed by Hope. After the idea had been defeated the matter was allowed to drop. Hope failed to find another assistant to teach the subject and with his advancing years (and "love of ease" according to his biographer Traill) the teaching of practical chemistry was abandoned and left to private lecturers. It was probably Hope's failure to encourage practical work and research from 1823 which led to the decline of Edinburgh from its leading position in Europe as a centre for the study of chemistry. Christison wrote pertinently "Dr Hope, with all his ability as a teacher, made very few chemists, because he never encouraged practical study."[54] Numbers of students registering for Hope's class declined gradually from their peak of 559 in the session 1823-24 to 188 in his final teaching session, 1842-43.

However much Hope may have been responsible for failing to stimulate interest in practical chemistry it is clear that in the first half of the nineteenth century there was a steadily increasing demand for chemical apparatus in Edinburgh. Those requiring such products were the chemistry professor and his assistants, extra-mural teachers, students, manufacturing chemists and those druggists who produced pharmaceutical products on a moderately large scale. Hope's own collection of apparatus was extensive. William Allen, who visited him in 1818, commented "Dr Hope called and took me to see his lecture room, which seems admirably contrived, and capable of holding several hundred pupils; in the apparatus room he has a considerable number of very large closets, in which the apparatus is nicely arranged for every lecture; and he has paged for each lecture with instructions to his assistant."[55] It is clear from the notes from which Hope lectured that his demonstration apparatus must have been extensive. Apart from Thomson, Fyfe, Anderson and Reid who have already been mentioned there were many others who lectured privately in chemistry who would require demonstration apparatus and some who would additionally have needed facilities for practical classes. Among the more prominent may be mentioned William Gregory (later himself professor of chemistry at Edinburgh), Edward Turner (later professor of chemistry at University College, London), George Wilson (later professor of technology at Edinburgh and first Director of the Industrial Museum of Scotland) and Kenneth Treasurer Kemp (a particularly inventive manufacturer of scientific apparatus). The extent of a lecturer's equipment can be gauged from a syllabus published by Reid[56] and from an engraving showing his laboratory at Roxburgh Place[57]; both indicate that large collections were required. The cost of setting up a chemistry laboratory might range from about £1000 (the sum spent during the first three years on Hope's new laboratory in the College) to £56 (the sum spent by Gregory during his first year as a private lecturer, see chapter four). Chemical works and pharmacies would also require supplies of chemical apparatus. An example of the former was the Magnesia Chemical Works at Bonnington, near Edinburgh, set up by Joseph Astley and later operated by his son Thomas[58]; John Anderson himself had been a manufacturing chemist there before becoming Hope's assistant. Large scale producers of pharmaceutical products in Edinburgh included John Fletcher Macfarlan and his partner David Rennie Brown, and Thomas and Henry Smith.[59] Both were involved in chloroform and alkaloid manufacture in the 1830s.[60] Some students of chemistry were expected to acquire a basic kit of chemical apparatus. Reid wrote in 1833 that "students . . . perform almost all the more important and interesting experiments, in an apparatus so small and portable, that a large proportion of the students of Practical Chemistry now provide themselves with it."[61]

The sources from which Hope acquired his extensive collection of apparatus are largely unknown, though he initially purchased Black's equipment (see chapter five). Evidence regarding the availability of scientific instruments from local makers and dealers is contradictory though there were no dealers in Edinburgh who traded on the scale of many of those in London.[62] A publication of 1825, referring to Hope's apparatus, jibed "Indeed, with all the well devised apparatus . . . and the facility with which any additional instrument can be procured in this city [Edinburgh] it might be expected that the professor of chemistry in this university would feel an ambition to excel in the excellence of his course of instruction."[63] On the other hand when Hope gave evidence to the Commissioners for visiting the Universities of Scotland he implied that he was supplied by London dealers, as well as confirming that he was personally responsible for the provision of apparatus:

> [Question by Commisioner] "Is there no allowance in your class for providing apparatus for the class? —
> [Hope] No, none.
> Is there a very extensive apparatus necessary for your class? —
> Certainly.
> Is any allowance given from any quarter whatever for the purpose of repairing such apparatus? — None.
> Looking to the present state of Science, and supposing a new Professor were called upon to teach that class, would it be a very serious and a very heavy outlay on his part to obtain an adequate apparatus? —
> It would require a considerable outlay.

Supposing a new Professor were to be appointed towards the commencement of a Session, could he possibly teach his class that year with any kind of success without the aid of your apparatus being transferred to him in some way or other? — My apparatus would certainly be of great service, yet the Science is so generally studied, that from London or Paris, especially from London, he would be supplied with apparatus sufficient to carry on his Winter course successfully, in the course of a few weeks".[64]

There are few clues as to which manufacturers or dealers Hope patronised. No account books or receipts survive among his notes in Edinburgh University Library. The only signed item in the Playfair Collection is the pyrometer by John Dunn who made instruments in Edinburgh from 1824 to 1841 (for further details, see catalogue object 7). Hope made a note on a slip of paper dated 1827 "Get from Dun what remains to compleat Air Weighing Flask".[65] In a bundle of notes describing safety valves produced by Robert Hare is a trade card for P and J Nimmo, Brassfounders.[66] It is quite possible that Hope specially commissioned pieces of apparatus from such local craftsmen, partly because he needed much of his demonstration apparatus to be constructed on a larger scale than usual so that it could be seen by his huge classes. Allen commented that "the bulbs of his differential thermometer are nearly two inches in diameter, and his apparatus, in general, is on a large scale . . . he has a glass receiver, capable of holding about ten or twelve gallons of oxygen".[67] The somewhat crude appearance of many of the items in the Playfair Collection which can be associated with Hope confirms that they were probably not constructed by specialised instrument makers. One local firm whose products did match the quality of those produced by the best London makers was Alexander Adie (after 1834, Alexander Adie and Son). A double register thermometer (now lost) was once in the Playfair Collection (see Appendix 1, item ii) and a balance which probably was Hope's property (see chapter five) were both constructed by Adie. For his chemical glassware, it is likely that Hope patronised the Edinburgh and Leith Glass-house Company (as Black had probably done) and the glassworks at Glasgow. Silliman recorded a visit he made with Hope in 1805: "To Dr Hope I was indebted for other civilities, — particularly in walking with me to Leith, to use his personal influence in obtaining for me some articles of glass apparatus, especially some instruments like those which I had seen successfully used in his own experiments."[68] Furthermore, among the notes of Hope survives a memorandum that the following items should be purchased: "Flasks from Leith or Glasgow, for the Expt of Uα [?] metals, as those of Cannongate melt too easily".[69] It is probable that Reid, when acting as Hope's assistant, acquired glass apparatus from Messrs Bailey and Company, of the Mid-Lothian Flint-Glass Works, Portobello.[70]

It might be expected that Hope's lecture notes on chemical apparatus would provide clues about his own equipment and its sources. Unfortunately few details are revealed, the lectures providing only a broad outline based on Black's treatment of the subject. What is of interest, however, is that Hope commended Lavoisier's treatment given in the *Traité élémentaire de Chimie* [71]:

"I am to make you acquainted with the Apparatus necessary for Chemical Expts ie that is the various instruments, vessels & utensils requisite for the Chemist — This part of my subject (not so engaging) yet abundantly useful — I shall treat it in a general way having no intention to detain you with the Description of all the complicated sorts of App contrived for particular Expts. In Mr Lavrs Excellent Work you have the most Ample Account of the App & figures of every article — consult it".[72]

Sadly Hope himself did not have his demonstration apparatus constructed to the same superb quality as that of the chemist whose treatise he was recommending.

Shortly after Hope started teaching in Edinburgh he published two papers, both of which involved the development of new pieces of apparatus. The first, in 1803, was the description of a new form of eudiometer which made provision for the absorption of gases by liquids. Hope demonstrated the apparatus in his lecture course and later sent a description to William Nicholson in London who published a description of it.[73] The second apparatus was that developed to demonstrate the temperature at which water attains its maximum density.[74] In its original form it consisted of a cylindrical glass jar around the middle of which a basin of tin was fixed for containing ice. Thermometers suspended near the top and bottom of the vessel recorded the temperature of the water. Hope demonstrated that the maximum density of water occurs at 39.5°F. In a number of other experiments described in a note to this paper Hope showed that the suggestion of Count Rumford (Benjamin Thompson) that liquids are non-conductors of heat was false.[75] For this his apparatus consisted simply of two concentric cylindrical vessels of tinned iron, the smaller suspended in the larger. The outer vessel had a gutter running round at the height of the base of the inner vessel; this provision was to allow for a flow of cold water, thus preventing heat being conducted through the sides of the apparatus. Hot water was poured into the inner vessel and a thermometer was placed near the bottom of the outer vessel which contained cold water. The rise in temperature indicated the flow of heat through a liquid. Hope presented a paper describing these experiments to the Royal Society of Edinburgh in January 1804. The apparatus used to determine the temperature of the maximum density of water became to be known, and still is referred to, as 'Hope's Apparatus'. The original instruments have not survived. However, George Wilson, Director of the Industrial Museum of Scotland in 1858 when the Playfair Collection was acquired, recorded in his annual report that "One of the thermometers employed by the late Dr Hope in determining the maximum density of water" had been obtained.[76] Unfortunately there is no record in the Museum Register of this being the case and the instrument cannot be traced.

Hope wrote no scientific papers between 1804 and 1838 apart from a joint report with Thomas Telford on the water supply of Edinburgh, published in 1813, for which he conducted a number of chemical analyses.[77] Towards the end of his life, however, he produced five papers though none of these proved to be of great consequence. The first was read to the Royal Society of Edinburgh in January and March 1836 and dealt with the colouring matter of plants which were used as indicators in titrations.[78] Hope's thesis was that most of these plants contain two agents which react, one with an acid, the other with an alkali, to produce the distinctive colours. The range of experiments in this paper is large and indicate that Hope had maintained an interest in botany. This was followed up seven years later by further experiments to test the effect of alkali on three plants.[79] Here he congratulated himself on his prowess in organic chemistry and the excellence of his lectures, being reported thus:

> "the author [Hope] described some remarkable properties . . . which have escaped the observation of all the investigators of organic substances, and some of which he had been in the habit of exhibiting in his lectures for nearly half a century".[80]

In April 1838 Hope read a paper to the Royal Society of Edinburgh on the temperature at which sea water attains its maximum density, continuing the experiments which he had described 34 years previously and using similar apparatus.[81] Traill claimed that he had prompted Hope to undertake this investigation.[82] Two minor papers completed this burst of activity in Hope's old age: a paper on inorganic nomenclature was read in April 1836[83], his final contribution being a speculation explaining a climatological phenomenon, 'The Freezing Cavern of Orenburg'[84] which had been observed and reported by the President of the Geological Society of London, Sir Roderick Murchiston. This latter paper was read only one month before Hope's death.

There is evidence that Hope was rather embarrassed at his poor record of research and publication. Certainly his colleague Traill was critical of this aspect of his work when in Hope's obituary he remarked "It is true that it is the paramount duty of one appointed to teach a science to make that his principal object; but this, I humbly conceive, is quite consistent with the most extensive original research".[85] Hope excused his unproductiveness in a remarkable written apologia which came into Traill's possession:

> "Those who devote themselves to the science of chemistry, may be divided into two classes — 1st, Those whose labours are employed in original researches, to extend our knowledge of the facts and principles of the science. 2dly, Of those whose business it is, from university or other appointments, to collect the knowledge of all that has been discovered, or is going forward in the science, to digest and arrange that knowledge into lectures, to contrive appropriate and illustrative experiments, and devise suitable apparatus for the purpose of communicating a knowledge of chemistry to the rising generation, or others who may desire to obtain it. From my professional situation, I consider myself, as Dr BLACK had done before me, as belonging to the second class of chemists. I consider my vocation to be the teaching the science".[86]

It was certainly as a teacher of chemistry that Hope's contemporary reputation lay. In addition to his university teaching, he was involved in popularising chemistry to a keenly interested public. In 1800 he gave a summer course in chemistry at the request of a number of lawyers from the Faculty of Advocates.[87] Hope was involved in setting up the Society of Arts for Scotland and is listed as an extraordinary councillor on the first printed syllabus of June 1821.[88] The aims of the Society were similar to those of its London counterpart, to provide artisans with an opportunity to keep abreast of developments in science and technology and to allow them to present their own ideas of discoveries, inventions and improvements in the 'useful arts', for which prizes might be awarded. The Society appears to have attempted to promote a type of mechanics institute in 1820; this was not effected but in August 1822 Hope was elected to a committee to "prepare and carry on the necessary measures" to establish permanent Schools of Arts in Edinburgh and other towns.[89] Later in the same month Andrew Fyfe was asked to deliver a course of lectures on the "Chemical Arts" but barely a week later a poorly attended meeting of the Society's Council (at which Hope was present) reversed the decision.[90] At a reorganisation of the Society in 1828 Hope was again appointed a councillor.[91] Another society with which Hope was connected was the Royal Medical Society, a body composed of undergraduate medical students. He was first elected to the Society in 1786;

50 years later he was appointed to a committee set up to make arrangements for celebrating its centenary. At the anniversary dinner it was Hope who occupied the Chair.[92]

Hope's most ambitious venture into extra-mural education was his 'Popular Lectures on Chemistry' which commenced in February 1826, the course of which was delivered three times a week running through to April. Hope gave 22 lectures of about one and a half hours' duration which were designed to cover 'The General Principles and more important facts of CHEMISTRY'.[93] He demonstrated his more spectacular experiments to an audience which included women; Henry Cockburn maliciously reported "the ladies declare that there never was anything so delightful as these chemical flirtations. The Doctor is in absolute extacy with his audience of veils and feathers, and can't leave the Affinities."[94] Nevertheless the course seems to have covered many of the more important aspects of heat, pneumatic chemistry, acids, salts and electricity.[95] A second similar series of lectures was delivered in 1828 but no further courses appear to have been given. In what possibly was a weak moment Hope offered the proceeds of these lectures (which cost two guineas per person) to establish a prize for the University students of chemistry. He may not have predicted how lucrative the course would prove, for in 1828 Hope presented the Senatus Academicus the sum of £800 (according to Cockburn, reluctantly). Hope's prize was offered for the best preparations by students of inorganic or organic compounds: in 1835 competitors were required to submit compounds of iodine and bromine, in 1837 isomorphous crystals and in 1840 "Compounds of Carbon & Hydrogen (not gaseous)". In setting the 1837 prize, Hope said to his class: "Though one Individual only can be successful in obtaining the Medal, but every competitor gains a prize of mugh higher value, He acquires that improvement in his knowledge in the Science & in his ability for experimental research, which it is my object to encourage."[96] Comparing this attitude with his previous behaviour Hope betrays a strange ambivalence towards practical chemistry, reinforced by the final words of the peroration to his course:

> "You are now well aware that Chemistry is an Experimental Science & will readily be convinced that it cannot be pursued with success by study in the closet alone —
> It is necessary to engage in the labors of the Laboratory — Experiments must be performed & processes conducted."[97]

Just before the commencement of the 1843-44 session Hope, with little warning, resigned his chair. He was then 77 years of age and his health had started to deteriorate, though unlike Black he had delivered his lecture courses up to this time without an assistant to aid him. There being insufficient time to elect a successor before the start of the session his colleague Thomas Stewart Traill, professor of medical jurisprudence, was asked to step in to give the annual course using Hope's lecture notes and demonstration apparatus. It is interesting that Traill commented that though most of Hope's material was up to date, his notes on organic chemistry required "various alterations and additions.[98] Soon after Traill had completed the course, in June 1844, Hope died. As might be expected he had accumulated some considerable wealth; the value of his estate totalled £55,000.[99]

NOTES AND REFERENCES

1. The earliest and still the most comprehensive biography of Hope is the obituary written by his colleague Thomas Stewart Traill, professor of medical jurisprudence: 'Memoir of Dr Thomas Charles Hope' *Trans Royal Soc Edinburgh* **16** 419 (1849). Other accounts rely heavily on this, for instance James Kendall, 'Thomas Charles Hope M.D.' in A Kent (ed) *An Eighteenth Century Lectureship in Chemistry* (Glasgow 1950) 156 and R H Cragg 'Thomas Charles Hope (1766-1844)' *Medical History* **11** 186 (1967).

2. Traill *ibid* 419 mentions that Hope went to school in Dumfries in 1778 but there appear to be no records of which school and no reason is given for why he was sent there.

3. Edinburgh University Library MSS Dc.10.9-15.

4. Thomas Charles Hope *De Plantarum Motibus et Vita Complectens* (Edinburgh 1787). The copy in the British Library is inscribed "Sir Jos. Banks Bt With the most respectful Compts. of his obedt. Servt. Thos. C. Hope".

5. Glasgow University Archives MS GUA 26693, University Minutes volume 78, f256 (entry for 10 October 1787).

6. *Ibid* f304 (entry for 16 May 1788), f313 (entry for 10 June 1788).

7. This comment is made by Sir James Hall in a letter to his uncle, William Hall, of 30 October 1788, see V A Eyles 'The Evolution of a Chemist' *Annals of Science* **19** 176 (1963).

8. Glasgow University Archives MS GUA 26694, University Minutes volume 79, f18 (entry for 20 July 1789), f19 (entry for 27 July 1789), f21 (entry for 23 October 1789, Hope's commission), f23 (entry for 27 October 1789, submission of a dissertation by Hope, *De Respiratione* to prove his abilities in medicine). The commission is dated at the Court of St James, 18 July 1789, two days before Stevenson made his request to the Senate.

9. Thomas Charles Hope 'Postscript to the History [of the Society]' *Trans Royal Soc Edinburgh* 3 141 (*recte* 143) (1794), Hope mentions that the mineral was brought to Edinburgh "in considerable quantity".

10. Edinburgh University Library MS Gen 48D (lecture notes taken during the 1796-97 session).

11. Thomas Charles Hope 'Account of a Mineral from Strontian, and of a Peculiar Species of Earth which it Contains' *Trans Royal Soc Edinburgh* 4 3 (1798).

12. This matter was first raised in detail by Thomas Thomson while reviewing Robert Jameson's *A System of Mineralogy* in *Annals of Philosophy* 8 133 (1816). Recent contributions have been J R Partington 'The Early History of Strontium' *Annals of Science* 5 157 (1942) and Andrew Kent 'Thomas Charles Hope and Strontium' *Actes du XIIe Congrès International d'Histoire des Sciences* 6 (Paris 1971) 45.

13. Adair Crawford 'On the Medicinal Properties of the Muriated Barytes' *Medical Communications* 2 301 (1790).

14. Letter from Adair Crawford to Andrew Duncan *Medical Commentaries* 4 433 (1789).

15. Hope *op cit* (11) 19 (footnote). 'Mr ASH' is probably Edward Ashe who matriculated in the medical faculty of Edinburgh University in 1786 and again (as Edward Ash) in 1787. He must have been a fellow student of Hope though he did not graduate. Ash was a President of the Royal Medical Society in 1788, see James Gray *History of the Royal Medical Society* (Edinburgh 1952) 81, 317.

16. Glasgow University Archives MS GUA 26695, University Minutes volume 80,f52 (entry for 26 October 1795): "The Faculty are truly Sensible of the loss they sustain by Dr Hope's Resignation. His diligence and success first as Lecturer in Chymistry, and then as Professor of Medicine did honour to himself and the Society, and his candour and probity of Manners recommended him to the esteem and friendship of his Colleagues."

17. Traill *op cit* (1) 422. Traill implies that Black gave the first lectures of the 1796-97 course, but reference to notes taken during this session reveals that Hope gave the first three lectures, after which Black lectured on heat (lectures 4-21) and 'Earths' (41-52), see Edinburgh University Library MS Gen 48D.

18. D L Cowen 'The Edinburgh Pharmacopoeia' in R G W Anderson and A D C Simpson (eds) *The Early Years of the Edinburgh Medical School* (Edinburgh 1976) 34. Some of the chemical processes in 1817 edition were criticised in a review in *Annals of Philosophy* (New Series) 1 58 (1821) by the editor, Richard Phillips. Hope vigorously defended himself of this attack, *ibid* 187. For a comment on this dispute, see Andrew Duncan junior *Edinburgh New Dispensatory* eleventh edition (Edinburgh 1826) 525.

19. Royal Infirmary of Edinburgh Archives MS Managers Minute Book volume 6, f337 (entry for 6 January 1800); Hope was appointed manager on the first possible occasion after Black's death.

20. *Proc Royal Soc Edinburgh* 1 114 (1845).

21. George P Fisher *Life of Benjamin Silliman M.D., LL.D.* 1 (London 1866) 165.

22. Robert Christison *The Life of Sir Robert Christison, Bart.* 1 (Edinburgh and London 1885) 57.

23. F Darwin (ed) *The Life and Letters of Charles Darwin* 1 (London 1887) 36.

24. Edinburgh University Library MSS Gen 268-272. About 170 envelopes survive, though these include notes used for teaching medicine at Glasgow, notes of popular lectures and bundles of redundant notes. Hope's annual course comprised 130-140 lectures.

25. These statistics can be found in *Evidence, Oral and Documentary taken by . . . the Commissioners . . . for Visiting the Universities of Scotland. Volume 1. University of Edinburgh* (London 1837) *Appendix* 130. Some caution is needed in drawing conclusions from numbers registering, as students could attend lectures given by extra-mural teachers. The professor of anatomy, Alexander Monro *tertius,* was not a popular lecturer and some students went elsewhere for anatomical instruction.

26. Fisher *op cit* (21) 164.

27. Edinburgh District Council Archives, MS Town Council Record volume 132, f232 (entry for 15 January 1800).

28. *Ibid* volume 137, f314 (entry for 27 October 1802).

29. *Ibid* volume 156, f292 (entry for 22 August 1810).

30. *Ibid* volume 157, f209 (entry for 5 December 1810).

31. *Ibid* volume 160, f388 (entry for 6 May 1812).

32. *Ibid* volume 161, f146 (entry for 8 July 1812).

33. D B Horn *A Short History of the University of Edinburgh* (Edinburgh 1967) 121.

34. Copies of this circular (dated 22 July 1815) and replies to it by the professors (some scribbled on the circular itself, others written out in the form of a letter) can be found in a bound volume of manuscript and printed material relating to the rebuilding of the College, Edinburgh University Library, Df.4.47.

35. This document has no title (other than the heading *Trustees for Rebuilding the University of Edinburgh*), no date and no place of publication; presumably it was circulated amongst the Faculty for discussion. A copy is to be found in the volume referred to in ref (34). The way in which Hope's report is set out (with what are presumably the architect's recommendations) is confusing.

36. *Ibid* 4. Amongst Hope's papers (Edinburgh University Library MS Gen 271) a plan of a laboratory survives, in his hand. This shows a series of seven rooms (labelled 'Cleaning', 'Materials & App', 'Crystallering' [sic], 'Dry Room', 'Retiring Room' and 'Blowpipe'; one, the largest, has no name but it is probably the classroom) for chemistry and additionally incorporates a two-roomed porter's house. Dimensions are given of some rooms. Whether this was intended as a guide for Playfair, whether it represented the 1791-1823 laboratory or whether it was a conjectural sketch of an ideal teaching laboratory is not clear.

37. William Henry Playfair *Report Respecting the Mode of Completing the New Buildings for the College of Edinburgh* (Edinburgh 1816).
38. Edinburgh University Library, drawings signed by W H Playfair, numbers 3, 5, 6, 8, 9, 10, showing details of the accommodation provided for chemistry teaching.
39. Christison *op cit* (22) 57.
40. Fisher *op cit* (21) 164.
41. Edinburgh University Library MSS Gen 270 and 271, two notebooks referred to as 'List of Specimens'.
42. *Ibid* Gen 271, see sheets headed 'Apparatus to be got this Summer – 1827', 'MM for Apps. 2P. Oct 24' and 'Facienda 1831-32'.
43. Christison *op cit* (22) 58.
44. The attitude of Hope towards teaching practical chemistry has been dealt with in some detail, see J B Morrell 'Practical Chemistry in the University of Edinburgh, 1799-1843' *Ambix 16* 66 (1969).
45. Christison *op cit* (22) 58.
46. *Ibid* 59. Although Christison indicates that his student society was a flourishing one, when giving evidence to the Commission visiting the Universities of Scotland in 1826 he stated "it is impossible for a student to pursue Practical Chemistry alone; the expense of procuring apparatus is enormous." (see *Evidence, Oral and Documentary op cit* (25) 292).
47. Edinburgh University Library MS Minutes of the Senatus Academicus volume 3, f377 (entry for 1 May 1823).
48. David Boswell Reid *Remarks on Dr Hope's "Summary", Presented to the Patrons of the University* (Edinburgh 1833) 9.
49. Reid had previously studied chemistry under Hope and Anderson. He also gained practical experience at the Magnesia Chemical Works at Bonnington; Reid wrote "Under the late Joseph Astley [the proprietor of the Works] I studied Chemistry both theoretically and practically. I was engaged as assistant in his extensive manufactories, and lived for a long time in one of his establishments, where I was constantly engaged in conducting experiments and processes upon the large scale, and in his experimental laboratory". see *Testimonials Regarding Dr D.B. Reid's Qualifications as a Lecturer on Chemistry and as a Teacher of Practical Chemistry* (Edinburgh 1833) 64.
50. A committee appointed in 1828 made a number of recommendations for extending the duration of existing courses and requiring additional courses. The proposal that a course of practical chemistry be established was amended by the College Council to include pharmacy, and it was resolved "That a Course of Practical Chemistry and Pharmacy be required of not less than three months' duration, the number of pupils in each class being limited to Twenty five." Royal College of Surgeons of Edinburgh, MS Minute Book of the Incorporation of Surgeons volume 10 (1828-32) f210, meeting of 1 June 1829.
51. Edinburgh University Library MS Gen 271 bundle 118 'Introductory Practical Course'. Hope's notes state "The Courses of Practical Chemistry in this University 1. Three months' duration 2. One hour every day but Saturday 3. No. in each class limited to 25" and "In all these duties, the whole labour is devolved on Dr Reid – He is well fitted for the Task by his thorough knowledge of the Subject – his Zeal, and ample and successful experience –".
52. Reid *op cit* (48) 4.
53. For bibliographical details of these pamphlets, see Morrell *op cit* (44) 74, note 43.
54. Christison *op cit* (22) 62.
55. *Life of William Allen, with Selections from his Correspondence 1* (London 1846) 355.
56. David Boswell Reid *Remarks on the Present State of Practical Chemistry and Pharmacy* (Edinburgh 1838) 12.
57. Frontispiece to David Boswell Reid *Textbook for Students of Chemistry* (Edinburgh 1836), engraving by John Johnstone.
58. *Testimonials Regarding Dr D.B. Reid's Qualifications op cit.* (49) 41.
59. Deric Bolton 'The Development of Alkaloid Manufacture in Edinburgh' *Chemistry and Industry* p.701 (1976).
60. G Colman Green 'William Gregory M.D., F.C.S.' *Nature 157* 467 (1946).
61. David Boswell Reid *A Memorial to the Patrons of the University on the Present State of Practical Chemistry* (Edinburgh 1833) 13.
62. For a general discussion on the possible suppliers of scientific apparatus in Edinburgh, see chapter five.
63. *The Contrast: or, Scotland as it was in the Year 1745; and Scotland in the year 1819* (London 1825) 189.
64. *Evidence, Oral and Documentary op cit* (25) 279.
65. Edinburgh University Library MS Gen 271, sheet labelled 'Apparatus to be got this Summer – 1827'.
66. *Ibid* MS Gen 268, bundle 32. Hare suggested the first of a number of proposals for oxy-hydrogen blowpipes, see 'Memoir on the Supply and Application of the Blowpipe' *Phil Mag 14* 238, 298 (1802).
67. Allen *op cit* (55) 355. The diameter of bulb recommended by John Leslie who arranged the manufacture of differential thermometers in Edinburgh, was $\frac{4}{10}$ to $\frac{7}{10}$ inch, see John Leslie *An Experimental Inquiry into the Nature and Propagation of Heat* (Edinburgh 1804) 10, and see catalogue section, objects 10 and 11. For a description of Hope's gas reservoirs, see David Boswell Reid *Elements of Chemistry* third edition (Edinburgh 1839) 793.
68. Fisher *op cit* (21) 161.

69. Edinburgh University Library MS Gen 271, sheet labelled 'MM for Apps. 2P. Oct 24'. For details of the Glasgow works, see catalogue section, objects 40-63, ref (10). The 'Cannongate' glass probably refers to that produced by William Ford from about 1810 in the North Back Canongate, to the west of Holyrood Palace, Edinburgh. These works were leased in 1819 by Bailey and Company, the arrangement continuing until 1835. See Revel Oddy 'Scottish Glass Houses' *The Circle of Glass Collectors Paper 151* (1966) 5 and ref (70) below.

70. *Testimonials Regarding Dr D.B.Reid op cit* (49) 38.

71. A L Lavoisier devoted more than one-thirds of his *Traité élémentaire de Chimie* (Paris 1789) to a discussion of chemical apparatus, illustrated by thirteen splendid plates.

72. Edinburgh University Library MS Gen 271 bundle 119 'Apparatus – for Chemical action'.

73. 'Account of a Simple Eudiometric Apparatus Constructed and Used by Dr T. C. Hope' *Nicholson's Journal 6* 61, 210 (1803). There is no example of a Hope eudiometer in the Playfair Collection but four specimens survive in the Museum of the History of Science, Oxford, see C R Hill *Museum of the History of Science Catalogue 1: Chemical Apparatus* (Oxford 1971) 27, items 113-116.

74. Thomas Charles Hope, 'Experiments and Observations upon the Contraction of Water by Heat at Low Temperatures' *Trans Royal Soc Edinburgh 5* 379 (1805).

75. Benjamin Thompson 'New Experiments upon Heat' *Phil Trans Royal Soc 76* 273 (1786), 'Experiments upon Heat' *Phil Trans Royal Soc 82* 48 (1792).

76. *Sixth Report of the Science and Art Department to the Committee of Council on Education* (London 1859) Appendix F, 102. Fairly early examples of Hope's apparatus survive at the Science Museum, London (inv no. 1857-47) by Hachette et Cie, Paris (see J A Chaldecott *Heat and Cold* (London 1954) 10, item 14); and at Transylvania College, Lexington, Kentucky, see Leland A Brown *Early Philosophical Apparatus* (Lexington 1959) 44.

77. Thomas Telford and Thomas Charles Hope *Reports on the Means of Improving the Supply of Water to the City of Edinburgh, and on the Quality of the Different Springs* (Edinburgh 1813).

78. Thomas Charles Hope 'Observations and Experiments on the Coloured and Colourable Matters in the Leaves and Flowers of Plants, Particularly in Reference to the Principles upon which Acids or Alkalies Act in Producing Red and Yellow or Green' *Proc Royal Soc Edinburgh 1* 126 (1845).

79. Thomas Charles Hope 'Chemical Observations on the Flowers of the Camellia Japonica, Magnolia Grandiflora, and Chrysanthemum Leucanthemum, and on three Proximate Principles which they Contain – Conclusion' *Proc Royal Soc Edinburgh 1* 403, 419 (1845).

80. *Ibid* 421.

81. Thomas Charles Hope 'Inquiry whether Sea-Water has its Maximum Density a Few Degrees above its Freezing Point, as Pure Water has' *Trans Royal Soc Edinburgh 14* 242 (1840).

82. Traill *op cit* (1) 429.

83. Thomas Charles Hope 'Observations on the Chemical Nomenclature of Inorganic Compounds' *Proc Royal Soc Edinburgh 1* 139 (1845).

84. Thomas Charles Hope 'An Attempt to Explain the Phenomena of the Freezing Cavern at Orenburg' *Proc Royal Soc Edinburgh 1* 429 (1845).

85. Traill *op cit* (1) 431.

86. *Ibid*.

87. *Ibid* 423. This seems to have been an informal arrangement, there being no evidence among the documents in the Advocate's Library that the lectures were organised officially.

88. National Library of Scotland MS Dep 230, item 71 (2), first circular of the Society, June 1821. Hope was voted on to a committee "on objects connected with the chemical arts" on 18 November 1822, see *ibid* item 11 'Minute Book of the Council of the Society of Arts for Scotland' volume 1.

89. National Library of Scotland MS Dep 230, item 1 'Minute Book of the Society of Arts for Scotland' volume 1 (9 July 1822 – 6 January 1830), Minutes of a committee meeting held on 27 August 1822.

90. *Ibid*. Minutes of committee meetings held on 30 August and 5 September 1822.

91. *Ibid*. Minutes of a committee meeting held on 3 December 1828.

92. Gray *op cit* (15) 188.

93. See advertisements in *The Caledonian Mercury* 28 January, 4 February and 13 February 1826; also in the *Edinburgh Evening Courant* 4 February, 11 February, 13 February 1826.

94. *Letters Chiefly Connected with the Affairs of Scotland, from Henry Cockburn to Thomas Francis Kennedy* (London 1874) 137; see also J B Morrell 'Science and Scottish University Reform: Edinburgh in 1826' *British J Hist Sci 6* 55 (1972).

95. Edinburgh University Library MS Gen 271 bundle 124 'Progress Popular Lect 1826/& 1828'.

96. Edinburgh University Library MS Gen 270 envelope 113.

97. *Ibid*.

98. Traill *op cit* (1) 433.

99. Scottish Record Office MS 14281/3 (SC 70/1/66).

CHAPTER 4

WILLIAM GREGORY

William Gregory differed in a number of respects from those who had preceded him in the chair of chemistry at Edinburgh. He did not spend the major part of his career as an Edinburgh professor, being appointed at the relatively late age of 41. Unlike his predecessors he pursued chemical research for much of his life, publishing frequently; he also wrote text books. He was the first to be appointed to a chair of chemistry at Edinburgh rather than a chair of medicine and chemistry.

William Gregory was born in December 1803 into a family with a long academic tradition, his great great grandfather being James Gregory, professor of mathematics at St Andrew's in 1669 and at Edinburgh in 1674.[1] His grandfather, John Gregory, was appointed to the Edinburgh chair of the practice of medicine in 1766 and his father, James Gregory, was professor of the institutes of medicine from 1776 and of the practice of medicine from 1790. In 1814 at the youthful age of nearly eleven William Gregory entered Edinburgh University to attend the arts class, and he matriculated into this class for seven consecutive years, transferring to the medical classes thereafter. He attended the chemistry lectures of Thomas Charles Hope for the first time in 1820-21 and repeated this course during the two subsequent sessions. In 1824 Edward Turner, an Edinburgh medical graduate, started giving private classes in chemistry in Edinburgh and Gregory acted as his assistant for two years, helping Turner with his research into the chemical analysis of micas, for which he received a flattering acknowledgement in the resulting published paper.[2] In the winter of 1827-28 Gregory travelled to Paris to study pharmaceutical chemistry with Pierre Jean Robiquet.[3] He returned to Edinburgh in the summer of 1828 and graduated MD in July, presenting the thesis *De Principiis Vegetabilium Alkalinis*.

In 1827 Turner was appointed the first professor of chemistry at the newly founded University College, London. Teaching started in the following year and he asked Gregory to join him as his assistant. Gregory was entrusted with teaching practical chemistry, the first such course to be offered in London.[4] Once again he helped Turner with his private research. In 1829 Gregory returned to Edinburgh to set up as a private lecturer in chemistry. Courses given by extra-mural teachers were recognised by the University, and private teaching was quite a normal start to an academic career in science or medicine in Scotland. Gregory's account book in which he recorded the expenses involved in installing himself as a private lecturer survives, which provides an interesting insight into what materials and apparatus were required.[5] Assuming that he recorded all his outgoings the sum spent amounted to £56·5s 7½d for his first year of teaching. The subheads under which he recorded his costs are: books and printing, brass apparatus, furnaces and tin apparatus [nil], glass apparatus, porcelain apparatus and earthenware, miscellaneous, laboratory and classroom, chemicals and specimens, platinum and silver apparatus, hardware, and clothes (no entry for the first year, but in February 1831 the item "a brown flowered silk dress waistcoat" appears). In only one instance is a scientific instrument maker or dealer named, 'Dunn', who supplied Gregory with two brass stands with transferrable rings.[6] Three textbooks were bought in 1829 and 1830, these being Michael Faraday's *Chemical Manipulation*, David Boswell Reid's *Elements of Practical Chemistry* and John George Children's *The Use of the Blowpipe in Chemical Analysis*.

Gregory's course was designed mainly for students of medicine. The University required that a six month course of theoretical chemistry be undertaken before graduation in medicine was permitted, and in 1829 the Royal College of Surgeons of Edinburgh decreed that candidates for its diploma had to have attended a three month course in practical chemistry. Gregory's notes for his theoretical chemistry course have been preserved.[7] These include the outlines of 27 lectures (presumably only the first part of his course) on chemical affinity, heat and the properties of gases. Lecture demonstrations are included, these being numbered up to 142. The experiments undertaken indicate that Gregory had a wider range of instruments available than is indicated by the purchases recorded in his account book. Later courses are also outlined in the same volume of lecture notes, and indicate that Gregory started his practical course towards the end of the first university session in which he lectured. These are headed 'Practical Chemistry commenced Feby 1st 1830' and 'Course of Practical Chemistry. Spring 1832'. The former series outlines 29 experiments, dated in such a manner to suggest that each experiment was performed by two groups of students at an interval of a week, presumably because Gregory's laboratory was not large enough to accommodate all his pupils at once. The latter series offered only seven experiments and was possibly part of a 'popular' course in chemistry. Yet another lecture series which Gregory delivered was a specialised course in organic chemistry.[8] This was offered each Summer from 1831 to 1835 and was the first such instruction available in Britain (Gregory claimed France and Germany as well).

While operating as a private lecturer in chemistry, Gregory contributed to the Hope-Reid debate on the establishment of an independent teaching part in practical chemistry at the University (see chapter three). In 1834 he published a pamphlet in which he supported Hope's view that the proposed lectureship should be created under the control of the professor of chemistry, thereby not infringing his professorial privileges.[9] In this publication Gregory outlined the aim of his own course: "Our business is what the medical boards require. While this is fully provided for, the teacher, if required, can give private instructions to the few who wish to pursue Practical Chemistry farther; but for such pupils the practical courses required by the College of Surgeons are not the proper field." After Reid left his employment as Hope's assistant he returned to private teaching and indulged in price-cutting tactics much to the annoyance of his colleagues, including Gregory who was forced to reduce his fee for theoretical and practical courses from six to four guineas. It was probably this "insufficient encouragement" financially (James Forbes' words) which led to him abandoning his private lecturing in 1835.[10]

Gregory started his independent chemical research while operating as a private lecturer in Edinburgh. Throughout his career he published about 65 papers of a scientific nature, mostly on chemical topics.[11] The first of these was one of his most important, 'On a Process for Preparing Economically the Muriate of Morphia', published in 1831.[12] The method Gregory evolved was to precipitate the morphine base with ammonia, add hydrochloric acid, and crystallise out the morphine as the hydrochloride. Two years after publication of the method the firm of J F Macfarlan started commercial production of morphine by Gregory's method and Edinburgh became an important centre for the manufacture of alkaloids.[13] Gregory's former teacher Robiquet discovered codeine as an impurity in morphia salts prepared by his pupil's method in 1832.[14]

In 1835 Gregory returned to the Continent to visit Justus Liebig at Giessen in Germany to learn the "new methods of organic analysis".[15] Liebig had been appointed professor of chemistry in 1824 and his research school had already acquired a reputation, being attended by many who were to become foremost chemists. It was during this visit that Gregory undertook his research into the properties of the liquid obtained by the destructive distillation of caoutchouc (in which he must have obtained isoprene)[16] and into the reaction between ammonia and sulphur dichloride, discovering nitrogen tetrasulphide. Results of the latter experiment were presented by Gregory at the *Versammlung deutscher Naturforscher und Ärzte* held at Bonn in September 1835, which impressed Jöns Jacob Berzelius, the highly respected Swedish chemist.[17] Gregory struck up a lifelong friendship with Liebig.[18] He translated seven of Liebig's works into English, the first being *Instructions for Chemical Analysis of Organic Bodies* of 1839.[19]

Having abandoned his self employment as a private chemistry lecturer at Edinburgh, Gregory took up his first appointment in 1836 at the Park Street School of Medicine, Dublin, a private medical school which had been set up in 1824.[20] As well as teaching, Gregory collaborated with James Apjohn (professor of chemistry at the Royal College of Surgeons, Dublin) in an investigation of the properties of 'eblanine', a substance obtained by the distillation of wood.[21] Apjohn considered Gregory to be an "able and highly instructive lecturer".[22] However Gregory was to stay in Ireland for only one session of teaching. In February 1837 Edward Turner died and Thomas Graham, professor of chemistry at Anderson's College, Glasgow, was appointed to succeed him at University College, London. Gregory applied for the vacant Glasgow post, and aided by an excellent testimonial by Liebig, was almost unanimously elected to the chair. Of the period of teaching which followed, Gregory later wrote: "[I] taught large classes, both theoretical and practical, at the Andersonian University during the sessions 1837-38 and 1838-39. I also gave a popular course, as I had frequently done in Edinburgh; and I lectured, besides, in Glasgow to a large mechanics' class."[23]

In February 1838 James Bannerman, 'Mediciner' (a post which at the time was equivalent to that of professor of medicine and chemistry) at King's College, Aberdeen, died. He had held the appointment since 1813, treating it as a sinecure and refusing to teach, even though he was pressed to do so by the Senatus from 1824. It was decided to delay advertising for a replacement because of the possibility that King's College and Marishal College (which functioned as independent universities in Aberdeen) might be merged as a result of a recommendation by the Royal Commission on Scottish Universities. However Thomas Clark, professor of chemistry at Marischal College, tried unofficially to persuade Thomas Graham to apply for the post. As Graham would not allow his name to be put forward, Clark approached Gregory. In October 1838 the Senatus asked the Chancellor, Lord Aberdeen, to proceed with the replacement of Bannerman, and in January 1839 he indicated that as proposals for the union of the universities were at an end, he would arrange for the appointment to be made. He then asked the Senatus to send him the names and testimonials of the applicants (there were at least three). In February the Senatus indicated that the Chancellor should not consider himself restrained to select from those who had applied, but that he should feel free to appoint the most deserving candidate known to him. Thus on 15 February 1839 Gregory was selected as Mediciner.[24]

Although for tuition in most subjects King's College and Marischal College offered quite separate courses, there had existed a joint medical school from 1818.[25] However in April 1839 the arrangement between the Colleges was dissolved and a house was purchased at Kingsland Place, Broadford (near the Infirmary) for the purpose of setting up an independent King's College medical school. Gregory, who had just arrived, was closely involved in instituting this new establishment. He did not, however, teach chemistry during his first session at Aberdeen. Chemistry classes had been held since 1817 by Patrick Forbes, professor of humanity (chemistry was a compulsory subject at King's College for arts students who wished to take the degree of master of arts). Gregory decided to allow Forbes to continue to teach chemistry during the 1839-40 session while he taught materia medica at the new medical school[26] though in the Summer of 1840 he gave a practical course in chemistry. This arrangement with Forbes could not continue, however, for in 1839 the Royal College of Surgeons of Edinburgh ruled that courses undertaken by medical students wishing to qualify for its diploma had to be taught by professors or lecturers who lectured in one subject only during any given session. This disqualified Forbes' course and he devoted his energies henceforth to teaching Latin. From October 1840 the Senatus required the Mediciner to teach chemistry on five days of the week for a six month session. From this time the post became to be known as the chair of medicine and chemistry.

In Gregory's first chemistry class for the arts course which started in 1840, 36 students registered. Gregory proved to be a popular lecturer and by 1843 his class numbered 74 students.[27] The average numbers of medical students during his period of teaching at Aberdeen was ten. Gregory's lecture course changed radically from his earlier teaching at Edinburgh: the so-called imponderables, heat, light, electricity and magnetism were omitted (being left for the natural philosophy classes), historical discussion was nearly totally neglected and the organic chemistry content was increased.[28] It was partly to improve his knowledge of the last-mentioned topic that Gregory again visited Liebig at Giessen in the Summer of 1841 where he intended to learn the latest improvements in organic analysis. Liebig had been provided with a new laboratory in 1839, built with State aid. It may have been Gregory's visit to these fine facilities which prompted him to write an important memorandum to the Earl of Aberdeen on the value of the teaching of chemistry and the benefits of the German system of the States financing the establishment of laboratories for training in practical chemistry. This was published in 1842 as *On the State of the Schools of Chemistry in the United Kingdom*. In this pamphlet Gregory stressed the importance of chemistry to the economy of the nation:

> "if any nation is bound to encourage and promote the study of practical chemistry, it is the British nation, which has derived, and continues to derive, such vast advantages from the applications of its principles to the useful arts. Yet, if we investigate the subject, we shall find that the opportunities afforded in this country for the study of practical chemistry are exceedingly limited;"[29]

Gregory then went into some detail about ideal facilities for practical chemistry:

> "There are required, first, a convenient and spacious laboratory, expressly fitted up for the purpose; secondly a complete apparatus; thirdly a large supply of fuel; fourthly, the substances or materials without which chemistry cannot be practised; and lastly, a qualified assistant, capable of taking charge of the laboratory, and of superintending under the professor, the working pupils; besides preparing the experiments for the lectures, which, for the benefit of beginners, should always be given in a practical school ...
>
> With regard to apparatus, besides the common and indispensable articles of glass, porcelain, and metal, including portable furnaces for charcoal and gas, and spirit lamps, every laboratory should possess one or more delicate but strong balances, with accurately divided weights; one or more air pumps ...
>
> A store of glass and other ordinary apparatus should be kept in a distinct room, which should be sold at prime cost to the students, that each may have his own, as far as the smaller articles are concerned ...
>
> A small room, clear of the laboratory, is also required for the balances air-pumps, and other delicate apparatus, which would be injured by corrosive vapours. In this room, a select library of chemistry ought to be kept ... and a cabinet of chemical specimens ...
>
> I regret to say, that hardly any university in this country, and but few on the Continent, can be said to possess one half of this necessary accommodation.
>
> In the laboratory of the University of Giessen, built and furnished under the superintendence of Professor Liebig, all the conveniences above mentioned, and a good many more, have been supplied."[30]

Gregory developed the argument that laboratories should not be the responsibility of the university and the apparatus that of the professor ("In most cases, the apparatus, or a great part of it, has been purchased by the professor, and consequently belongs to him, an arrangement fraught with inconvenience in the case of death or removal of the professor, and the appointment of his successor"). He then described the financing of Liebig's laboratory at Giessen: "the laboratory has been built, on plans approved by him [Liebig], at the expense of the government; and the entire cost, including all furnaces, sand-baths, water-pipes, and the numerous indispensable fixtures of a laboratory, amount to 13,000 florins, or about £1,120 sterling." Gregory also criticised the duty which had to be paid on the import of chemical glass and porcelain imported from the Continent ("The heavy duty (or rather duties, for there are two, amounting to about 50 or 60 per cent) on foreign glass is a most serious obstacle to the cultivation of chemistry in this country.") finally he summed up the situation in British universities:

"I do not know what means the richer universities of Oxford, Cambridge, and Dublin, in none of which does a school of research at present exist, may have of establishing such a school; but the Scottish universities have no funds applicable for this purpose. All of them have some locality set aside as a laboratory, more or less complete in several cases; but in two, namely King's College, Aberdeen, and St Andrew's, very deficient. In King's College, the lecture-room, a very good one, is the only laboratory; and it is quite unfit for the accommodation of students engaged in research."[31]

Gregory was not totally discouraged by the lack of government finance for science, however, and he made the best of his limited resources, improvising where necessary. The professor of natural philosophy at King's College, John Fleming, wrote a testimonial which praised Gregory's encouragement to his students;

"your manner renders you of easy access to the student, and your feelings dispose you cheerfully to facilitate his progress. I may add, that your power of contrivance or adaptation enables you to conduct many processes with a few pieces of common or cheap apparatus, and thus encourages the student, by demonstrating that much may be done in the science of Chemistry with very scanty resources."[32]

While in Aberdeen Gregory continued to pursue his research and to publish his work. In 1840 at the British Association for the Advancement of Science in Glasgow he presented a number of results including an investigation of the derivatives of uric acid.[33] In 1842 he emphasised again his interest in the organic chemistry of natural products by translating Liebig's *Die Thierchemie oder die organische Chemie in ihrer Anwendung auf Physiologie und Pathologie*.[34] Together with Liebig and Wilson Turner, Gregory edited the sixth edition of Edward Turner's *Elements of Chemistry* which was also published in the same year.[35] Gregory was a founder member and the first secretary of the Aberdeen Philosophical Society which first met in January 1840.[36] He read several chemical papers at meetings of the Society, discussing Liebig's work on a number of occasions.

In 1843 Thomas Charles Hope resigned from the chair of medicine and chemistry at Edinburgh. Though he was 77 years of age his decision was without warning and there was no possibility of appointing a successor to teach the chemistry class for the 1843-44 session. Thomas Stewart Traill, professor of medical jurisprudence, was asked to give the chemistry lectures and he agreed. The Edinburgh chair was Gregory's ultimate ambition. Initially many possible candidates were mentioned but the main protagonists narrowed down to Gregory, Andrew Fyfe and Samuel Brown (a controversial chemist who claimed to have performed experiments in transmutation). Gregory's published testimonials formed an impressive list of the leading chemists of the time, including Thomas Thomson, Thomas Graham, Justus von Liebig, Friedrich Wöhler, Robert Bunsen and Jöns Jacob Berzelius.[37] Brown withdrew before the day of election and Gregory was elected with twenty votes to Fyfe's fourteen on 14 May 1844. In ratifying the appointment the Senatus, anxious to avoid a fiasco of the type encountered with Hope and Reid over the teaching of practical chemistry (see chapter three), declared that the new Professor "shall not only give regular course of lectures during the ordinary sessions of the said College or University, but shall also teach the Science of chemistry practically to the students' attending therein."[38]

One of Gregory's first actions as professor at Edinburgh was to make changes to his laboratory accommodation. In October 1844 his classroom and the laboratory were inspected by the College Committee of the Town Council and certain works were recommended such as the fitting of blinds, the boarding of the stone floor and other minor matters. As almost a symbolic gesture Gregory specially requested "that the present furnaces should be removed and two frames fitted up, each containing three of Liebig's furnaces."[39] Surprisingly, in view of his publication of 1842 concerning ideal facilities for practical chemistry teaching, Gregory made no further request to the Town Council during his period of office other than for the provision of a personal retiring room.[40] His successor found the laboratory to be quite inadequate for his teaching.

Figure 9 William Gregory (1803-1858), lithograph by F Schenk after W Stewart, ca 1845. (Reproduced by permission of Edinburgh University Library)

Shortly after coming to Edinburgh Gregory started to write his chemistry textbooks. The first, *Outlines of Chemistry for the Use of Students,* was published in 1845.[41] In the preface to this work he explained that he was excluding topics such as heat, light, electricity and magnetism, which though they had been taught in the past, he did not consider them to be chemical subjects.[42] The *Outlines* was published in a second edition in 1847, later editions appearing under two separate titles. The first of these, *A Handbook of Organic Chemistry,* was published in 1852.[43], the other part, *Handbook of Inorganic Chemistry,* being published in the following year.[44] The former work was the first specialised textbook on organic chemistry written by a British chemist. In its preface, Gregory referred to "the unusually favourable reception" which his *Outlines* had received "it having been adopted as a text book in the Universities of Oxford, Cambridge, and Dublin, besides many other schools". The *Handbook of Organic Chemistry* had the distinction of being published in a German translation.[45] Gregory's last textbook, the *Elementary Treatise on Chemistry* of 1855 was originally written for the eighth edition of the *Encyclopaedia Britannica.*[46] The profusion of books which Gregory published at Edinburgh may account for the fact that no copies of student notes of his lectures appear to have survived.[47]

At Edinburgh his own research continued. One of his most widely appreciated contributions was his paper of 1851, 'Notes on the Purification and Properties of Chloroform'[48] James Young Simpson, professor of midwifery at Edinburgh, introduced chloroform as an anaesthetic in 1847 and by the following year it was being produced by at least three firms of manufacturing chemists in quantity. However when wood spirit was used in its production impurities remained which could cause vomiting and nausea in the patient. Gregory and Alexander Kemp (of Kemp and Company, an instrument making firm which made among other things, ether inhalers) devised a method by which a physician could purify his own chloroform by washing it with sulphuric acid, followed by treatment with manganese dioxide to remove any sulphurous acid which was formed. This was his final chemical paper, the last twelve papers published between 1853 and 1857 dealing with microscopic observations of diatoms. Several of these were presented to the Royal Society of Edinburgh. Gregory had been elected a fellow in February 1832 and from November 1844 until his death he was annually elected an Ordinary Secretary of the Society.

Another activity which occupied Gregory during his second Edinburgh period was the translation of more of Liebig's books into English. In 1847 he published an edition of *Chemische Untersuchung über das Fleisch*[49] and a fourth English edition (with Lyon Playfair) of *Die Chemie in ihrer Anwendung auf Agricultur und Physiologie.*[50] In the next year a translation of *Untersuchen über einige Ursachen der Saftebewegung im Thierischen Organismus*[51] appeared, followed in 1855 by *Die Grundsätze der Agricultur-Chemie.*[52] In 1851 Gregory edited the third edition of Liebig's popular *Familiar Letters on Chemistry.* The letters had first appeared in the *Augsburg Allgemeine Zeitung* between 1841 and 1844 and the first collected edition of them was in fact an English translation of 1843. In the preface to the fourth edition, written a few months after Gregory's death, Liebig paid Gregory the following generous tribute: "My friend Dr Gregory, who assisted me in the former editions, has, meantime, been removed by a Higher Power from his family and from science. United for more than twenty years by the ties of a sincere and intimate friendship, no one feels more keenly than I do, the great loss sustained by his friends in his death. Rare extent of knowledge, combined with admirable powers of mind, enabled him to apprehend exactly the ideas and views of others, and to put them forward in the best form, and with the precision of original composition."[53] It was partly due to Gregory that Liebig's work was appreciated in Britain and it was partly the reputation of Liebig and his research school that promoted the state financing of scientific education in Britain.

Gregory's fame in Edinburgh in the late 1850s did not rest solely on his teaching, research, translations and scientific writings. He was deeply interested in and closely associated with a number of pseudo-sciences such as phrenology, clairvoyance, animal magnetism (and its development mesmerism).[54] He translated two of Carl von Reichenbach's books, in 1846 *Abstract of "Researches on Magnetism and on Certain Allied Subjects" including a Supposed new Imponderable* and in 1851 *Researches on Magnetism, Electricity, Heat, Light, Crystallisation, and Chemical Attraction, in their Relation to the Vital Force.* Also in 1851 he published a book of his own on the subject, *Letters to a Candid Enquirer on Animal Magnetism,* produced as well in an American edition in the same year. This work was published into the present century, the fifth edition appearing in 1909. Gregory's obsession did not pass uncriticised by the academic community. Allen Thomson, professor of anatomy at Glasgow University wrote a blistering review of *Letters to a Candid Enquirer:*

> "[They] give an elaborate description and analysis of the phenomena of animal magnetism, which he [Gregory] considers to be identical with mesmerism, electro-biology, electro-psychology, and hypnotism. We need scarcely say that Dr Gregory regards all the objections raised against his favourite science as illogical and absurd in the extreme; that he admits, almost without question, all the lower as well as the higher phenomena; that clairvoyance, pre-vision, retro-vision, intro-vision, and transference of the senses, present to him no insurmoutable difficulties. He is an enthusiastic advocate of, and a firm believer

in animal magnetism, under whatever name it is presented to him. Being more of a chemist than a physician, he was less qualified than many other members of his profession to form sound opinions, and a correct judgment, respecting doctrines and practices, the just appreciation of which required no small amount of physiological and psychological knowledge."[55]

Gregory was a staunch supporter of the Phrenological Society of Edinburgh; the *Phrenological Journal* and *The Zoist* contain many papers and other references to his contributions between 1832 and 1854.[56] Inevitably his excessive credulity brought much derision, especially when he participated in public revelations of his beliefs. His His colleague Robert Christison wrote a particularly contemptuous account of such an occasion in a letter of April 1856. It is in the same vein as Lord Cockburn's account of Hope's popular lectures (see chapter three) and deserves to be quoted at length:

"Ulbster Hall is the offspring of the conversion into shops of No 132 George Street, and occupies the entire drawing-room front of that home. Miss Catherine Sinclair has long looked daily from her house at the unoccupied room on the opposite side of the street with a painful sense that so good a room must positively be turned to some good account; and so she determined to improve the manners and customs of Edinburgh Society by substituting evening lectures for balls; and having rented the room, and christened it Ulbster Hall, she has already entertained no less than 150 fashionables, on each of twelve occasions, with music, tea, and coffee, and a lecture or reading of a play . . .

Dr Bennett gave an admirable lecture, under the odd title of the 'Physiology of Macbeth', upon the rational view to be taken of such of the mental phenomena of mesmerism as are true; and in the course of it he pounded satisfactorily Dr Gregory, who always sits in the front seat of every concert, opera, lecture, or public spectacle, but found his position a very hot and uneasy one for once in his life. This led to a sort of reply, in which he dosed his audience with all the common trash of mesmerism, clairvoyance, table-turning and spirit-rapping, and declared his belief in all of them and in every alleged fact connected with them. He had the egregious simplicity to declare that he had seen a table, of its own accord, making 'gracious movements', and walking from one part of the room to another; and that he believed, on the authority of a witness whose testimony was indisputable, that a pet table had in a similar fashion followed its mistress up-stairs like a dog. I was not present, thinking it a shame to encourage in any shape the Professor of Chemistry in making a donkey of himself and a laughing-stock of the University."[57]

The thaumaturgical device (or 'tetragrammaton') in the Playfair Collection (object 31) can almost certainly be associated with Gregory and was possibly used in clairvoyance activities. He expressed his belief in this subject in a letter written in December 1851 concerning a personal experience.[58] It is difficult to reconcile Gregory, the prolific and painstaking research chemist with Gregory, the gullible believer in the paranormal, though there are other examples of competent scientists of the period being drawn into such movements.

Gregory had been afflicted with painful illness from his youth and his decline was protracted. His practical classes, at least in the final years of his teaching, were conducted by his assistant, Allen Dalzell.[59] Gregory died in April 1858. Remarks made by George Wilson[60] and Lyon Playfair[61] after his death indicate that his teaching powers had declined in his last years. His colleague William Pulteney Alison, professor of medicine, said in his obituary "He had formed plans which, had his health permitted, would have resulted in a course of chemical instruction not surpassed in extent and importance in any school in Europe."[62] Whether these plans included the establishment of a research school is not clear. Certainly Gregory left no Giessen-like inheritance for his successor.

NOTES AND REFERENCES

1. The Gregory family and its remarkable academic record (in three generations of the family there were sixteen Gregories who occupied professional chairs in British Universities) has been investigated by Agnes Grainger Stewart *The Academic Gregories* (Edinburgh 1901) and John D Comrie *History of Scottish Medicine 1* (London 1932) 383. The most exhaustive treatment is by Paul David Lawrence 'The Gregory Family: A Biographical and Bibliographical Study' University of Aberdeen PhD thesis, July 1971 (unpublished); a genealogical table showing the professions of the Gregory family is included in volume one, opposite page 298. Biographical details of William Gregory may be found in an obituary by William Pulteney Alison 'Account of the Life and Labour of Dr William Gregory' *Proc Royal Soc Edinburgh 4* 122 (1857-62) and in G Colman Green 'William Gregory, M.D., F.C.S.: 1803-1858' *Nature 157* 468 (1946).

2. Edward Turner 'On Lithion-Mica' *Edinburgh J Sci 3* 261 (1825).

3. Turner himself had studied under Robiquet, as had Robert Christison, professor of medical jurisprudence; see Robert Christison *The Life of Sir Robert Christison, Bart. 1* (Edinburgh and London 1885) 267, for a description.

4. *Testimonials in favour of William Gregory M.D., F.R.S.E.* (Edinburgh 1853); this evidence appears in a testimonial offered by Robert Warrington, who at the time was a fellow assistant with Gregory to Turner.

5. Aberdeen University Library MS 2206/24 'Account of Expenses for Apparatus &c incurred in setting up as a Teacher of Chemistry by Dr William Gregory beginning Martinmas 1829 Edinburgh'.

6. *Ibid*, entry dated 1829. The maker referred to is almost certainly John Dunn who made the demonstration pyrometer in the Playfair Collection, see catalogue section, object 7.
It is interesting that Dunn supplied an air pump to University College, London, in December 1827, presumably ordered by Turner, see John Dunn 'Description of an Improved Air-Pump' *Edinburgh New Phil J 4* 382 (1828).

7. Aberdeen University Library MS 2206/25 'Outlines of Lectures on Chemistry Begun October 1828'.

8. Gregory *op cit* (4) 9.

9. William Gregory *Observations on the Proposed Appointment of a Teacher of Practical Chemistry in the University* (Edinburgh 1834); see also J B Morrell 'Practical Chemistry in the University of Edinburgh 1799-1843' *Ambix 16* 75 (1969).

10. Gregory *op cit* (4) 55.

11. The precise number of papers which Gregory published is not easy to estimate. The *Royal Society Catalogue of Scientific Papers 3* (London 1869) lists 53. Laurence *op cit* (1) *1* 221 cites 150, though many of these are repeated exactly or in editorially altered form (some translated) in different journals; others, notably those published in the *Phrenological Journal and Miscellany* are scarcely of a scientific nature.

12. William Gregory 'On a Process for Preparing Economically the Muriate of Morphia. With a letter from Dr Christison on its Employment in Medicine' *Edinburgh Med and Surgical J 35* 331 (1831).

13. Green *op cit* (1) 467. See also Deric Bolton 'The Development of Alkaloid Manufacture in Edinburgh 1832-1939' *Chemistry and Industry* p 701 (1976).

14. Pierre Jean Robiquet 'Nouvelles Observations sur les Principaux Produits de l'Opium' *Annales de Chimie 51* 225 (1832).

15. Gregory *op cit* (4) 11.

16. Gregory took some of the distillate with him to Giessen for Liebig's opinion; it had first been prepared by a Mr Enderby of London, see 'On a Volatile Liquid Procured from Caoutchouc by Destructive Distillation . . .' *London and Edinburgh Phil Mag* (New Series) *9* 321 (1836).

17. Although Gregory presented his results at Bonn (see Gregory *op cit* (4) 40, 59) he first mentioned the discovery at the Royal Society of Edinburgh in December 1835, see 'Notice of a New Compound of Sulphur, which is probably a Sulphuret of Nitrogen' *Proc Royal Soc Edinburgh 1* 106 (1845).

18. Justus Liebig *Familiar Letters on Chemistry, in its Relation to Physiology, Dietetics, Agriculture, Commerce and Political Economy* fourth edition (London 1859) vi.

19. Originally published as *Anleitung zur Analyse organischer Körper* (Braunschweig 1837). It is of interest that Gregory's edition was published by Richard Griffin and Son of Glasgow. This firm was by this date supplying chemical apparatus and it seems likely that certain items in the Playfair Collection (for instance items 19 and 22) may have been supplied by them. The publication establishes a definite link between Gregory and Richard Griffin and Son. For further details, see chapter five.

20. J D W Widdess *A History of the Royal College of Physicians of Ireland* (Edinburgh and London 1963) 163.

21. James Apjohn and William Gregory 'Examination of Eblanine' *Irish Acad Proc 1* 33 (1841).

22. Gregory *op cit* (4) 29.

23. *Ibid* 5.

24. An account of the curious and somewhat irregular selection of Gregory as Mediciner is told in greater detail in Alexander Findlay *The Teaching of Chemistry in the Universities of Aberdeen* (Aberdeen 1935) 49.

25. Comrie *op cit* (1) *1* 391; John Malcolm Bulloch *A History of the University of Aberdeen 1495-1895* (London 1895) 169.

26. *Aberdeen Journal* 23 October 1839.

27. Gregory *op cit* (4) 11.

28. Findlay *op cit* (24) 54.

29. William Gregory *Letter to the Right Honourable George, Earl of Aberdeen, KT . . . on the State of the Schools of Chemistry in the United Kingdom* (London 1842) 12.

30. *Ibid* 14. For details of state support for Liebig's laboratory, see J B Morrell 'The Chemistry Breeders: the Research Schools of Liebig and Thomas Thomson' *Ambix 19* 39 (1972).

31. *Ibid* 34. The first direct state-financed facilities for chemistry (other than those indirectly aided at the established universities) were provided at the Royal College of Chemistry in Oxford Street, London. This institution was privately financed from its foundation in 1845 until 1853 when it came under the aegis of the newly established Science and Art Department of the Board of Trade; see D S L Cardwell *The Organisation of Science in England* (London 1972) 86 *et seq*.

32. Gregory *op cit (4) 27*.

33. William Gregory 'On the Pre-existence of urea in uric acid' *Brit Ass Report 1840: Trans of the Sections* 73 (1841).

34. Justus von Liebig *Animal Chemistry, or Organic Chemistry in its Applications to Physiology and Pathology . . . Edited from the Author's Manuscript by W Gregory* London 1842).

35. Edward Turner *Elements of Chemistry, including the Recent Discoveries and Doctrines of the Science . . . Edited by J. Liebig, W. G. Turner, and W. Gregory* (London 1842).

36. *Trans Aberdeen Philosophical Soc 1* XIX (1840-84): "At a meeting held in Professor Gregory's house and attended by Professor's Fleming, Pirrie, Thomson and Gregory . . . It was agreed that a Society be constituted of the Cultivators of Science in and around Aberdeen".

37. Gregory *op cit* (4). It has sometimes been suggested that Lyon Playfair was a candidate for the Edinburgh chair of chemistry in 1844, even though he provided one of Gregory's testimonials. This confusion may arise from a letter written by Robert Christison to his twin brother Alexander on 29 June 1858 (published in Christison *op cit* (3) *1* 297) part of which reads "but as I told a town councillor of note - who consulted me in 1844, when Gregory and Samuel Brown were the favourites - that the Chair should be given neither to the one nor the other, but to young Playfair . . .".

38. Edinburgh University Library MS Minutes of the Senatus Academicus volume 7,f1 (entry for 4 October 1844).

39. Edinburgh District Council Archives, MS Town Council Record volume 242,f396 (entry for 22 October 1844).

40. *Ibid* volume 253,f95 (entry for 18 December 1849); volume 262,f87 (entry for 28 February 1854).

41. William Gregory *Outlines of Chemistry for the Use of Students* (London 1845).

42. *Ibid* v; see also Anon 'Obituary of Dr W. Gregory' *Edinburgh New Phil J 8* 173 (1858).

43. William Gregory *A Handbook of Organic Chemistry . . . Third Edition Corrected and Much Extended* (London 1852).

44. William Gregory *Handbook of Inorganic Chemistry; being a New and Greatly Enlarged Edition of the "Outlines of Inorganic Chemistry"; for the Use of Students. Third Edition, Corrected and Enlarged* (London 1853).

45. William Gregory *Gregory-Gerding's Organische Chemie . . . Selbständig bearbeitet von T Gerding* (Braunschweig 1854).

46. *The Encyclopaedia Britannica 6* eighth edition (Edinburgh 1854) 437.

47. I have recorded locations of approximately 55 sets of notes taken by Black's students and nine sets of Hope's lectures.

48. William Gregory 'Notes on the Purification and Properties of Chloroform' *Proc Royal Soc Edinburgh 2* 316 (1851).

49. William Gregory *Researches on the Chemistry of Food* (London 1847).

50. William Gregory and Lyon Playfair *Chemistry in its Applications to Agriculture and Physiology* fourth edition (London 1847).

51. William Gregory *Researches on the Motion of the Juices in the Animal Body* (London 1848).

52. William Gregory *Principles of Agricultural Chemistry* (London 1855).

53. Justus von Liebig *Familiar Letters on Chemistry in its Relation to Physiology, Dietetics, Agriculture, Commerce and Political Economy fourth edition* (London 1859) vi.

54. Phrenology, or the determination of an individual's mental faculties by means of studying the external conformation of the cranium, was enthusiastically followed in Edinburgh, the Phrenological Society being founded in February 1820 (see G N Cantor 'Phrenology in Early Nineteenth-Century Edinburgh: an Historiographical Discussion' *Annals of Science 32* 195 (1975)). In animal magnetism, a hypnotic state could be induced by an operator who was held to possess the power to emanate magnetic forces. A full discussion can be found in ref (55) below.

55. *The Encyclopaedia Britannica 20* eighth edition (Edinburgh 1860) 441. There is also a strongly critical review ('Odyle Mesmerism, Electrobiology, &c.') in *The British and Foreign Medico-Chirurgical Review 8* 378 (1851).

56. *Phrenological Journal and Miscellany 8* 161 (1832-34); *Phrenological Journal and Magazine of Moral Science 14* 22, 31, 37, 38, 49 (1842); *15* 38, 236 (1843); *17* 379 (1844); *18* 38, 54 (1845); *19* 123, 180 (1846); *20* 90 (1847); *The Zoist, A Journal of Cerebral Physiology and Mesmerism* contains further papers on the subject by Gregory: *4* 558 (1847); *5* 380 (1848); *9* 422 (1852); *10* 1, 399 (1853); *11* 349 (1854).

57. Christison *op cit* (3) *2* 328 (letter from Christison to 'D.C.', 7 April 1856). Christison and Gregory crossed swords at the British Association meeting at Edinburgh in 1850, Gregory maintaining that lead sulphite did not act as a poison, Christison rejecting his evidence, see *The Pharmaceutical J 10* 150 (1851).

58. Frank Podmore *Mesmerism and Christian Science* (London 1909) 189 (reprinted from *The Zoist 9* 422 (1852)).

59. After Gregory's death, Dalzell was paid the sum of £30 for the tables, shelves and other fixtures which he had installed in the laboratory (see Edinburgh District Council Archives MS Town Council Record volume 275,f198 (meeting of 3 August 1858)).

60. Jessie Aitken Wilson *Memoir of George Wilson* (Edinburgh 1860) 454.

61. Wemyss Reid *Memoirs and Correspondence of Lyon Playfair* (London 1899) 179.

62. Alison *op cit* (1) 122.

CHAPTER 5

PLAYFAIR COLLECTION: ACQUISITION AND PROVENANCE

William Gregory died on 24 April 1858. His successor, Lyon Playfair, was appointed professor of chemistry on 28 June of the same year. Playfair soon realised that Gregory's laboratory would be inadequate for his teaching; he later recalled "He [Gregory] had not been able to found a teaching laboratory on a scale commensurate with the importance of the chair. This I determined to do, so far as the limited accommodation then in the University would permit. Before entering upon my duties as a professor, a considerable sum had to be spent in equipping the laboratories and chemical museum with the full appliances for teaching."[1] Within a month of his appointment, Playfair had submitted to the College Committee of the Town Council a list of proposals of changes to his laboratory.[2] He recommended that a tiled demonstration bench be provided, that fume cupboards be reconstructed, that the laboratory be painted and that a number of other changes be made. At about the same time he started clearing his store-rooms of redundant apparatus, handing over 75 items considered to be of historic interest to his colleague George Wilson, Director of the recently established Industrial Museum of Scotland.[3] Thus the Playfair Collection of chemical apparatus was formed.

Wilson and Playfair may have met for the first time in Edinburgh in 1837 where they were fellow members of a student order of brotherhood. In the following year both became assistants to Thomas Graham, professor of chemistry at University College, London. In 1839 Wilson returned to Edinburgh to take his medical degree. Playfair ventured further, to Giessen in Germany, where he studied for a doctor of philosophy degree under Justus Liebig. Eventually he returned to England to teaching and in 1850 he was appointed Special Commissioner for the Great Exhibition which was being planned for the following year. Later, as Joint Secretary of the Science and Art Department of the Board of Trade, Playfair was closely involved in securing funds for the foundation in 1854 of the Industrial Museum of Scotland.[4] In February 1855 Wilson was appointed its first Director. His enthusiasm and his attitudes to public education made him a particularly inspired choice.[5] He received the additional appointment of professor of technology at Edinburgh University in August 1855. The duties associated with this post were to teach the principles of industrial science as illustrated by the Museum collections. His students, who numbered 40 in the session 1855-56, also studied chemical analysis in a laboratory attached to the Museum.

Wilson's earliest acquisitions for the Museum were mainly objects which illustrated the intermediate stages and final products of industrial processes, though in his second annual report Wilson commented "whilst seeking to avoid any intrusion on the domain of archaeology, the Director has solicited contributions of examples exhibiting the early forms of important instruments, and has begun by presenting an antique working model of the first glass friction electrical machine".[6] His interest in scientific instruments pre-dated his appointment as Director: in 1849 he had published an article on the historic development of the air pump.[7] Later he argued the case for the preservation of historic scientific and technological artefacts in his presidential address to the Royal Scottish Society of Arts in November 1857, when he stated as his acquisition policy:

> "The Industrial Museum will try to find space for the most elephantine machines; and as all gifts to it will be national property, formally catalogued, free to the public, yearly reported on, and officially inspected, it is likely as most institutions to preserve for the public the objects confided to its care. But at least let all those instruments which, as at once infants and parents, are especially memorable, be preserved."[8]

In fact Wilson's most significant acquisition of scientific instruments was the Playfair Collection which was recorded by him in the manuscript Museum Register on 30 September 1858. Before listing the items Wilson wrote as a heading "Specimens from the Laboratory of the Edinburgh University comprising Apparatus, Instruments etc used by Professors Black, Hope and Gregory". This seems to have been the first inventory to have been compiled of the apparatus and thus its certain provenance dates from this time.

It is not clear why Wilson was certain that the apparatus which he had acquired did not pre-date Black; the air-pump (object 1) was almost certainly constructed long before Black's appointment at Edinburgh and Wilson, with his interests in pumps, must have known this. He did not, however, extend the possible association to James Crawford, Andrew Plummer or William Cullen, who had occupied the Edinburgh chair from 1713 to 1766. It is possible of course that Wilson knew of evidence which eliminated any likelihood of its connection with Black's predecessors, but if he did, he did not reveal it. For this reason the chapters forming the introduction to the catalogue section have included discussion of the work and teaching of all the professors of chemistry at Edinburgh prior to 1858, and in the paragraphs which follow the continuity and transfer of the apparatus from Crawford up to Gregory will be considered.

Prior to the implementation of the Universities (Scotland) Act of 1858 the professors of chemistry were unsalaried.[9] Additionally they carried the financial burden of running their laboratories and supplying the apparatus and materials with which they taught. To compensate for this they received a substantial income from their students who paid a fee to attend a course of lectures (medical students were in fact required to have undertaken a course in chemistry before they could graduate). Since the apparatus was thus the property of the professor who acquired it, no official inventories were made and in theory the professor could dispose of it as he wished.[10] If any personal inventories were made by or for the professors of chemistry, none now seems to have survived. The only remaining documents which resemble in any way an inventory are the notebooks which Hope compiled for his assistant in which he listed the apparatus required for demonstrations to his class.[11] However the entries are too cryptic to enable the Playfair Collection instruments to be correlated with them. A further possible source of information regarding the apparatus, but one which proves disappointing, is the wills and inventories of the professors, all of which fail to mention any such possessions.[12] The implication is that arrangements were made concerning the disposal of apparatus (which could be of significant value) prior to the death of the owner and this possibility will be discussed below.

Nothing is known of the apparatus with which James Crawford may have taught. The first record of equipment acquired for the chemistry class is in the letter book of the partnership formed by the first few professors of the medical faculty, John Innes, Andrew Plummer, John Rutherford and Andrew St Clair (their purchases, laboratory and teaching activities have been mentioned in chapter one).[13] The apparatus, acquired over a period of seventeen years, became the property of Plummer when the partnership was finally dissolved in 1743. He continued teaching in his laboratory until the Summer of 1755 when serious illness prevented him from embarking on his next annual course. Among those proposed to replace him were Joseph Black, William Cullen and Francis Home. As Black was immediately available he started to teach, probably starting on 29 October, making use of Plummer's laboratory and apparatus. However on 19 November 1755 the Town Council appointed Cullen conjoint professor with Plummer, though without consulting him. This caused some resentment among members of the Senatus, and did not enable Cullen to use Plummer's teaching facilities. Though these problems were later resolved by the intervention of the Lord Provost, George Drummond, it is uncertain whether Cullen eventually acquired the laboratory of Plummer or that of James Scott (he may at one time have intended to set up his own facilities in the Infirmary; again these points are discussed in detail in chapter one). It is not known whether Plummer's apparatus was transferred, on his death in July 1756, to Cullen who became sole professor. Black took over the chemistry chair in November 1766 when Cullen was made professor of the institute of medicine. Though no details are known of the transfer of laboratory and apparatus, it is fair to assume that this was carried out in a cordial spirit and without the rancour of ten years previously. There is evidence that Black brought over from Glasgow some of the apparatus he had purchased from his own funds.[14]

It is certain that Hope started his lecture course in Edinburgh using Black's apparatus after he had been appointed joint professor in November 1795 (he probably purchased it between Autumn 1796 and December 1799 when Black died though no details of the transaction survive). The evidence for this was supplied by Hope when he was questioned by the Commissioners visiting the Scottish Universities in 1828 concerning the ease of purchase of demonstration apparatus:

> "[Commissioner] Did Dr Black leave any Apparatus?
> [Hope] Dr Black left a very excellent Apparatus, and a considerable collection of mineralogy, which, from a private arrangement, became my property."[15]

During Hope's 48 year occupancy of the Edinburgh chair he accrued a large quantity of apparatus which was admired by those who visited him - William Allen reported in 1818 that "in the apparatus room he [Hope] has a considerable number of very large closets, in which the apparatus is nicely arranged for every lecture."[16] In 1843 Hope resigned due to bad health and Thomas Stewart Traill, later professor of medical jurisprudence, stood in to conduct the 1843-44 session. Hope "freely gave" Traill "the use of his manuscript lectures, which were fairly written out, and of his whole apparatus".[17] William Gregory was elected professor of chemistry in May 1844 and Hope died a month later. The College Committee of the Town Council wrote "to the representatives of the late Dr Hope, requesting them to communicate with Dr Gregory in regard to the Apparatus belonging to them".[18] Unfortunately there is no further mention in the Council minutes of this matter and it is not known whether Gregory made arrangements to inherit his predecessor's apparatus.

It is possible that Gregory acquired apparatus from his own funds while teaching at King's College, Aberdeen. Unlike the situation at Edinburgh the professor at Aberdeen was not entirely responsible for providing his own equipment (Gregory's predecessor, Patrick Forbes, had purchased apparatus and chemicals from university funds to the sum of £210 7s 7d.[19]) though there are indications that Gregory had to improvise and that he possessed only "very scanty resources".[20] He had previously furnished his private laboratory at Edinburgh[21] and it is possible that he may have added to his collection in Dublin and Glasgow. All this may have been brought to Edinburgh in 1844 on his appointment as professor of chemistry. Gregory must have possessed an adequate research laboratory with which to pursue his own work: several of his research papers date from this last Edinburgh period. After Gregory's death, however, Lyon Playfair found the teaching laboratory to be inadequate.[1]

Figure 10　Lyon Playfair teaching at the Museum of Practical Geology, London, wood engraving from *The Illustrated London News*, 1852.
(Reproduced by permission of Edinburgh University Library)

Playfair presumably inherited the apparatus which had belonged to Gregory.[22] It may be assumed that those items presented to the Industrial Museum were no longer of value in the teaching activities of the laboratory. Indeed, much of the apparatus had probably been declared redundant: Playfair refers to it as having been "rescued from the lumber of the laboratory".[23] The 75 objects transferred must have formed a small fraction of the apparatus left on Gregory's death and must be considered to be a selection made by Playfair rather than a representative sample of what was present in the laboratory at the time. There appears to have been no discontinuity in the steady acquisition of laboratory apparatus from the time of Black (it does not seem likely that any of the professors removed all his equipment, for instance) and thus there is no reason why the apparatus discovered by Playfair should not have dated back to Black's accession to the Edinburgh chair, or indeed even earlier.

The possible source of individual items in the Playfair Collection is dealt with in the catalogue section but it is appropriate to discuss briefly how easily, and from where, apparatus could be obtained by the Edinburgh chemistry professors. Throughout the period London and Paris were main centres of the scientific instrument making trade in Europe, though it would appear from the general quality of the Playfair apparatus (and the absence of makers' signatures) that it was more likely to have been constructed locally than to have been made in the relatively sophisticated workshops further afield. A recent study of the trade in Scotland (up to 1900) describes it as "a provincial activity, serving a predominantly local market".[24] The sparsity of makers is evident when attempts are made to trace suppliers of chemical apparatus between 1766 and 1858, the likely period of acquisition of items forming the Playfair Collection. Between these years 163 instrument makers operating in Scotland have been recorded.[25] However this figure includes not only makers of chemical and philosophical instruments, but also barometer makers, clock, watch and chronometer makers, mathematical instrument makers, opticians, and makers of nautical and optical instruments. During the latter half of the 18th century only one chemical and one philosophical instrument maker in Edinburgh have been found in the Post Office and other Directories. For the first half of the 19th century, five chemical and fourteen philosophical instrument makers are recorded. Most of these were active only for brief periods, and their output is likely to have been small. Additionally, though, brassfounders, tinsmiths and other non-specialists may have produced instruments for the Edinburgh professors. It is recorded, for instance, that John Sibbald, described as a smith, made chemical furnaces for Black.[26]

The increasing demand for apparatus made by private lecturers and students of practical chemistry has been discussed in chapter three. David Boswell Reid, the lecturer who had at one time been Hope's assistant, was responsible for stimulating the supply of kits of chemical apparatus to students not only in Edinburgh but also in Montrose, Arbroath, and Dalkeith and possibly elsewhere. He gave courses of practical chemistry at schools (this must be one of the earliest examples of schoolboys studying the subject) and estimated in 1837 that over 400 students were undertaking tuition in the course which he had developed.[27]

Reid's *The Study of Chemistry: its Nature and Influence on the Progress of Society* published in about 1836, contains the advertisement "A complete assortment of Chemical Substances, with Apparatus sufficient to perform all kinds of Experiments, has been prepared and fitted up in an elegant and commodious manner, in a conveniently sized mahogany box, by J.F. MACFARLANE Chemist and Druggist, North Bridge, Edinburgh, and is sold for L.2,5s. Portable Laboratories arranged for Demonstrations in Schools and Academies, L.6,6s." It is not known from which source Macfarlane was obtaining his apparatus.

Two of the private lecturers were also suppliers but not, it seems, on a large scale. John Deuchar had premises in Lothian Street from 1815 to 1833[28], while Kenneth Treasurer Kemp founded the firm in 1828 which later became Kemp and Company.[29] John Dunn, who from 1833 until his death in 1841 was Curator of the Society of Arts for Scotland, started a small instrument firm in 1824 in West Bow, moving two years later to Thistle Street and in 1828 to Hanover Street. Hope is known to have patronised Dunn (see chapter four) and also the ingenious Kemp, from whom he bought electrical batteries.[30] An Edinburgh dealer known to have been involved in chemical apparatus was Alexander Allan, who in 1811 supplied St Andrew's University with a large quantity of apparatus which had previously been the property of Thomas Thomson.[31] John Miller (later in partnership with his nephew Alexander Adie) produced mathematical, optical and philosophical instruments of good quality in Edinburgh over the period 1774-1822. The firm also produced barometers, Adie patenting the sympiesometer in 1818.[32] A fine chemical balance by Adie (made in the period 1843-1877 when in partnership with his son) to a design of T C Robinson is known[33] but this appears to be an exceptional item and none of the other instrument makers so far mentioned specialised in chemical apparatus. Deuchar was a druggist while Kemp, Dunn and Allan probably dealt mainly in philosophical instruments. There was, however, one exception to this general situation in Scotland, the activities of the brothers Richard Thomas and John Joseph Griffin, who set up a specialist chemical apparatus making and dealing business in Glasgow. This may have started as early as 1826 though the earliest catalogue forms an appendix to the 8th edition of J J Griffin's *Chemical Recreations* and is dated 30 November 1837. J J Griffin had studied at the Mechanics' Institute in Glasgow (it is possible that he was taught by Gregory) at Paris and at Heidelberg (under Leopold Gmelin). As a result of his European travels he introduced new apparatus to the country: as one writer put it "Mr Griffin, to whom the chemical world is indebted for the introduction from the Continent of an immense number of instruments, until then unknown in England . . ."[34] Griffin mentioned in the Preface to *Chemical Recreations* that he had gathered apparatus and materials from France, Prussia, Bohemia, Austria, Saxony and Sweden. The range of instruments manufactured and sold by Richard Griffin and Company was extensive and the apparatus was relatively inexpensive; a set of instruments for elementary operations could be purchased for one guinea. Some of the stands and supports in the Playfair Collection were almost certainly supplied by the firm. The flourishing state of the business is indicated by the opening of a London branch by J J Griffin in about 1848 and the very comprehensive catalogue *Chemical Handicraft,* issued first in 1866, which includes nearly 5,000 entries.

In concluding this chapter it must be admitted that the pre-1858 provenance of the Playfair Collection is not well defined. Nevertheless as a collection which represents the equipment with which chemistry was taught, it is one of the few to survive from the period, and its associations with those teachers who were so influential in the development of the subject in British universities adds to its importance.

NOTES AND REFERENCES

1. Wemyss Reid *Memoirs and Correspondence of Lyon Playfair* (London 1899) 179.
2. Edinburgh District Council Archives MS Town Council Record volume 275,f196 (minutes of meeting of 3 August 1858).
3. The Industrial Museum of Scotland was founded in 1854 as a national museum. In 1864 the name of the Museum was changed to the Edinburgh Museum of Science and Art and in 1904 (fifty years after its foundation) the Museum was again renamed The Royal Scottish Museum.
4. Douglas A Allan 'The Royal Scottish Museum: General Survey' in *The Royal Scottish Museum 1854-1954* (Edinburgh 1954) 5.
5. Jessie Aitken Wilson *Memoir of George Wilson* (Edinburgh 1860)4.
6. Appendix C to the *Fourth Report of the Science and Art Department to the Committee of Council on Education* (London 1857) 166. This instrument, still in the collection of the Department of Technology, has the Register number 1856.562.
7. George Wilson 'On the Early History of the Air-Pump in England' *Edinburgh New Phil J 46* 330 (1849).
8. George Wilson 'Address as President of the Royal Society of Arts at its Annual Meeting, November 23, 1857' *Trans Royal Scottish Society of Arts 5* 43 (1861).
9. D B Horn 'The Universities (Scotland) Act of 1858' *University of Edinburgh J 19* 169 (1959).
10. The situation was different for the professor of natural philosophy, for whom equipment was purchased by the Town Council. The first record of this is of a grant of £50 offered in 1709 (see Edinburgh District Council Archives, MS Town Council Record volume 49,f371, meeting of 10 June 1709) for "procureing Instruments necessary for confirming and Illustrating by Experiment the truths advanced in the Mathematicks and Naturall Philosophy within the University". However inventories were not kept by the professors and the need for these was commented on by the Commissioners for Visiting the Scottish Universities in their report, see *Report made to His Majesty by a Royal Commission of Inquiry into the State of the Universities in Scotland* (London 1831) 81. From 1833 an inventory of apparatus in the Natural Philosophy Department was kept; this is now in the Royal Scottish Museum.
11. Edinburgh University Library MSS Gen 270 and 271, two notebooks referred to as 'List of Specimens'. These notebooks have been mentioned in chapter three and references to them are made in connection with a number of objects in the catalogue section.
12. Scottish Record Office MSS: Black (wills) 14281/2 (CC8/8/131/2), 1.4281 (RH 1/2/519); Hope (will) 14281/6 (SC 70/4/1), inventory 14281/3 (SC 70/1/66); Gregory (will) 14281/5 (SC 70/4/60), (inventory) 14281/4 (SC 70/1/106).
13. 'Books of Letters and Envoys Belonging to the Elaboratory'. Manuscript volume in private collection (Miss A Scott-Plummer).
14. Requests by Black for apparatus left behind at Glasgow to be sent on to Edinburgh can be found in letters from him to Watt, see Eric Robinson and Douglas McKie *Partners in Science* (London 1970) 8 (Black to Watt, 10 January 1768), 15 (Black to Watt, 28 February 1770) and 39 (Black to Watt, 22 May 1773). Black also claimed glassware from William Irvine, lecturer in chemistry at Glasgow, see chapter two, ref (44). The chemistry lecturers at Glasgow received an allowance for the purchase of apparatus which remained the property of the University, see chapter one.
15. *Evidence, Oral and Documentary taken by . . . the Commissioners . . . for Visiting the Universities of Scotland. Volume 1. University of Edinburgh* (London 1837) *Appendix* 283.
16. *Life of William Allen, with Selections from his Correspondence 1* (London 1846) 355.
17. Thomas Stewart Traill 'Memoir of Dr Thomas Charles Hope' *Trans Royal Soc Edinburgh 16* 419 (1849).
18. Edinburgh District Council Archives MS Town Council Record volume 242,f397 (meeting of 3 August 1858).
19. Alexander Findlay *The Teaching of Chemistry in the Universities of Aberdeen* (Aberdeen 1935) 46.
20. William Gregory *Testimonials in Favour of William Gregory M.D. F.R.S.E.* (Edinburgh 1853) 27 (this reference appears in a testimonial written by Gregory's colleague at Aberdeen, the Rev John Fleming, professor of natural philosophy).
21. Aberdeen University Library MS 2206/24 'Account of Expenses for Apparatus &c incurred in setting up as a Teacher of Chemistry' (covers period November 1829 to February 1831).
22. This is implied by George Wilson in his annual report for 1857, *op cit* (6), in which he says "pieces of scientific apparatus of historical, industrial interest have, through the co-operation of Dr Lyon Playfair, been obtained by the director from the collections of the late Dr Gregory".
23. Lyon Playfair *A Century of Chemistry in the University of Edinburgh: being the Introductory Lecture to the Course of Chemistry in 1858* (Edinburgh 1858) 13. It is possible that the apparatus which was no longer of use was kept in the lumber room attached to the chemistry laboratory. For its location, see chapter three.
24. D J Bryden *Scottish Scientific Instrument Makers 1600-1900* (Edinburgh 1972) 38.
25. *Ibid* 43-59.

26. *The New Dispensatory . . . by Gentlemen of the Faculty of Edinburgh* (Edinburgh 1786); see chapter two, ref (108) for details of Sibbald.

27. D B Reid 'Education - Study of Chemistry' *Chambers' Edinburgh Journal* 5 139 (1837).

28. Bryden *op cit* (24) 46. Robert Christison mentions in his autobiography that he and his friends who formed a chemical society "met first in an attic in my father's house in Argyll Square, and afterwards in a large underground room below the shop of one Deuchar, subsequently a lecturer on popular chemistry", see R Christison *The Life of Sir Robert Christison, Bart.1* (Edinburgh and London 1885) 59, 62.

29. Bryden *op cit* (24) 51. Kemp published a number of papers describing apparatus which he had invented, see for example 'Description of a New Kind of Galvanic Pile, and also of another Galvanic Apparatus in the form of a trough' *Edinburgh New Phil J 6* 70 (1829); Description of an improved Blowpipe' *ibid* 340; 'Description of two Thermometers' *Edinburgh J Nat Geogr Sci 1* 183 (1830). The Royal Scottish Museum possesses a number of items made by him: 1856.B1-B4, samples of liquefied gases; 1856.A1-A5, various batteries; 1856.C1, his improved blowpipe.

30. Edinburgh University Library MS Gen 271 folder 117.

31. Allan is recorded as having operated from Baron Grant's Close from 1806 to 1810 and from 9 Lothian Street from 1811 to 1835 (see Bryden *op cit* (24) 43); however a signed instrument, possibly a width gauge, at St Andrews University is dated 1804. The chemical apparatus bought by Robert Briggs, professor of chemistry at St Andrews, from Allan is listed in an inventory (St Andrews University Muniments SM110.MB F37). It was probably purchased from Thomson when he closed his practical chemistry class in Edinburgh and moved to London. The inventory may well define the apparatus necessary for this type of instruction at this date. The Department of Chemistry at St Andrews University possesses an interesting collection of early glass and ceramic ware but it is not possible to say whether this came to Briggs from Thomson via Allan.

32. Alexander Adie 'Sympiesometer, or Improved Air Barometer' British Patent 4323 (1818).

33. John T Stock and D J Bryden 'A Robinson Balance by Adie & Son of Edinburgh' *Technology and Culture 13* 44 (1972). See also Appendix 3, object 14, of this work.

34. C Greville Williams *Handbook for Chemical Manipulation* (London 1857) 159.

*Catalogue of the
Playfair Collection*

CONTENTS OF THE CATALOGUE SECTION

Of the 75 items of chemical apparatus acquired for the Industrial Museum of Scotland by George Wilson from Lyon Playfair in 1858, 64 survive. These extant items are listed in the Catalogue which follows; entries include object title, original description and number in the Museum Register (in square brackets), physical description, dimensions, illustration, brief history of the development of the class of object or instrument, and in certain cases associations with the teachers of chemistry and their work where this is possible. Notes on the items which cannot now be traced or which have been destroyed follow the Catalogue as Appendix 1. The Catalogue retains the order of listing adopted by George Wilson in the manuscript Museum Register: his original entries are reprinted as Appendix 2. A few objects in other collections can be associated with some certainty with the Edinburgh professors of chemistry: these are listed in Appendix 3.

The italicised letters which follow the objects in the contents list below indicate with which professor that item may first be associated. (Andrew Plummer *P*, William Cullen *C*, Joseph Black *B*, Thomas Charles Hope *H*, William Gregory *G*.) In several cases the association can be no more than tentative (or indeed speculative); for these a question-mark is included.

		Page
1.	Air pump *C, B?*	67
2.	Pneumatic trough *B*	71
3.	Balance *B*	73
4.	Lead block *B*	76
5.	Double bellows *B?*	77
6.	Wire cage *H*	80
7.	Pyrometer *H*	82
8.	Model of a bimetallic spiral thermometer *H*	84
9.	Air thermoscope *H?*	86
10.	Differential thermoscope ('pyroscope') *H*	88
11.	Differential thermoscope ('photoscope') *H*	88
12.	Voltaic pile *H*	91
13.	Sturgeon cell *H*	94
14.	Voltameter *H*	96
15.	Instantaneous light box *H*	99
16.	Instantaneous light box *H*	99
17.	Argand lamp *H?*	101
18.	Lamp furnace *B?*	103
19.	Retort stand, boss and three clamps *H,G?*	105
20.	Heating bath (?)	107
21.	Adjustable table support	108
22.	Multi-purpose apparatus support *H,G?*	109
23.	Still boiler *B?*	110
24.	Water-cooled condenser *B,H?*	112
25.	Counter-current condenser *H,G?*	112
26.	Iron bottle *B*	115
27.	Pattern for beehive shelves (?) *H,G?*	118
28, 29.	Two concave reflectors *H*	120
30.	Concave reflector *H*	120
31.	Thaumaturgical device *G*	123
32-35.	Four crucibles *G?*	125
36.	Series of absorption bottles *B?*	127
37.	Bottle with tap	129
38.	Graduated glass vessel *H?*	131

			Page
39.	Stoppered glass bottle	B?	133
40-42.	Three cucurbits	B?	135
43, 44	Two globular flasks	B?	136
45, 46	Two ellipsoidal flasks	H?	136
47.	Straight-sided flask	B, H?	136
48, 49	Two ovoidal flasks	B?	137
50.	Florentine flask	B?	137
51.	Flask with constricted neck	B?	137
52-54.	Three retorts	B?	138
55.	Cucurbit	H?	138
56.	Alembic	B?	139
57-60.	Four flasks	B,H?	140
61.	Bell-shaped vessel	B.H?	140
62, 63.	Two tubes for producing electrostatic charges	B	141
64.	Instantaneous light machine	H	148

APPENDIX 1 — 150

i.	Differential thermometer or thermoscope	H?	150
ii.	Double registering thermometer	H?	150
iii.	Stock of a pistol...	H	151
iv.	Experimental oil lamp		152
v.	Sectional model of a reverberatory furnace		152
vi-ix.	Wollaston's scales... 4 kinds	H	152
x.	Alembic		153
xi.	Flask or receiver		153

APPENDIX 2 — 154

APPENDIX 3 — 156

1.	Hydrostatic balance	P	156
2-7.	Six glass bottles	P	156
8.	Mortar and pestle	C	156
9.	Phial containing red powder	B	157
10.	Spirit thermometer	B?	157
11, 12.	Samples of barium hydroxide and strontium hydroxide	H	157
13.	Eudiometer	H?	157
14.	Chemical balance with weights	H	158

INSTRUMENTS AND APPARATUS IN THE PLAYFAIR COLLECTION

1. Hauksbee-type air pump
[Double barrelled Air Pump in the form in which it was first made by Hawksbee 1703-9. 1858.275.1]

Overall height 1054, frame base 380 x 380, cylinder length 209mm.

The framework and central decorative panel are carved from wood (probably pear). The pump cylinders, working mechanism, taps and table (on which a bell jar would be placed for experiments) are of brass. The panel (on which is carved a representation of a sunflower) conceals a gear wheel which operates two pistons by a rack and pinion mechanism (the piston packing has decayed). The axle of the gear, crank handle and two screws for securing the instrument to the floor are made of wrought iron. The glass barometer tube is a replacement. The number '111' is painted in white on the frame to the front of the instrument.

Figure 11   Hauksbee-type air pump

Figure 12   Detail of the carved panel which covers the mechanism

The air pump is supposed to have been first produced by Otto van Guericke in 1654.[1] It was improved later in the 17th century by Robert Boyle, Robert Hooke, Denis Papin and others.[2] On 15 December 1703 Francis Hauksbee *the elder* demonstrated an air pump to his design to the Royal Society in London (he was the Society's 'Operator' or 'Curator of Experiments') and he regularly used the air pump in his lecture demonstrations until his death in 1713.[3]

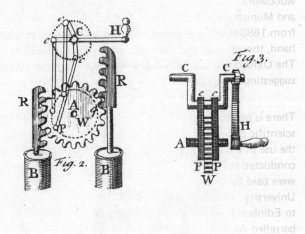

Figure 13 Rack and pinion mechanism; the path taken by the connecting rod of the former Vream mechanism can be seen carved into the wood

Figure 14 Vream mechanism, taken from J T Desaguliers *A Course of Experimental Philosophy 2* (London 1744) plate 24

Hauksbee fully described his pumps in his *Physico-Mechanical Experiments* of 1709[4]; they were said to resemble those produced by Papin in about 1675.[5] The double-barrelled pump incorporated a rack and pinion mechanism to raise and lower pistons: by turning the pinion by means of a handle, one piston was raised while the other was lowered. The direction of rotation was then reversed to complete the cycle. In 1717 William Vream (who had formerly been an apprentice to Hauksbee) suggested an improvement to Hauksbee's pump to allow continuous rotation of the handle in one direction instead of the reciprocating motion. This was achieved by incorporating a crank which was attached to the pinion by means of a connecting rod:

"Thus far Mr Hawksbee has describ'd the Pump; which I hope I have since improv'd by a Contrivance, whereby in turning the Winch quite round, the Emboli or pistons are alternately rais'd and depress'd;"[6]

This modification was also described and illustrated by J T Desaguliers in 1744 (fig 14):

"... the Racks are Shorter, and the wheel bigger than is usual in Hauksbee's Pumps."[7]

The Playfair Collection pump was almost certainly one of the Vream-variety, though in its present state it incorporates only the Hauksbee gearing. However there remain features which are different in a number of respects from the small number of Hauksbee-type pumps which survive.[8] The Playfair pump has a large housing for its remaining mechanism, and there is an unutilised hole at the centre of the sunflower carving (fig 12). This would have supported the crank, which would have been rotated by the handle. Moreover the brass pinion has a hole near its circumference which would have been connected to the crank by a rod. The ratio of gear teeth of the pinion to those on the racks is 2:1. This is a necessary consequence of the Vream mechanism. The other Hauksbee pumps all have much longer cylinders and a smaller ratio of gear teeth.[9]

Dating the Playfair Collection pump and the other Hauksbee pumps presents problems. The Vream pump was first described in 1717; a version of the Hauksbee free-standing pump was still available in 1766 when Benjamin Martin wrote:

> "... it [the air pump] received at last an elegant and magnificent Form from the ingenious Mr HAUKSBEE, and called it the Standing AIR-PUMP, which is yet in Use for those who will go to the Price of it."[10]

However by this latter date improved versions including a more compact form which could be used on a table were available.[11]

The design and decoration of the pumps can be of some value in dating them by comparison with furniture and other woodwork. The panel of the Playfair Collection pump indicates a date of ca 1710-20. The Royal Society, Longleat and Munich pumps appear earlier on stylistic grounds, the Royal Society example being earliest of all, possibly dating from 1680-90. This early date may possibly be misleading due to conservatism in design by its maker. On the other hand, though his pump was not exhibited until 1703, there is evidence that Hauksbee was working as early as 1680.[12] The Oxford example would seem to be the latest of the group, the carved design on the mechanism housing possibly suggesting a date as late as 1740.[13]

There is no direct evidence to associate the pump directly with either William Cullen or Joseph Black. Cullen's only scientific paper, 'Of the Cold Produced by Evaporating Fluids and of Some Other Means of Producing Cold', involved the use of a pump[14] though it was not described in his paper and in any case it seems likely that the experiments were conducted before Cullen left Glasgow for Edinburgh in 1755. Unlike the situation in Edinburgh, where instruments were paid for and were the property of the professor, at Glasgow most of the instruments were the property of the University. Thus while it is improbable that the pump which Cullen used was his own and that it travelled with him to Edinburgh (he may have used the pump described in a Glasgow University inventory of 1727 as "A large double barrelled Air Pump, with a Mercurial Gage and four Screws to fix it to the ground"[15]) it is known that he did buy some of his apparatus (see chapter one) and between 1736 and 1742 he enquired of "Mr Hawkby in Crane court Fleet Street" (ie Francis Hauksbee *the younger*) the exact price of "Cupping glasses with Air pu[mp]".[16]

Black himself described an experiment in which an air pump was used in his paper read in 1755 to the Philosophical Society of Edinburgh, 'Experiments on Magnesia Alba, Quicklime, and some other Alcaline Substances'.[17] However it is clear that he had difficulty in finding an efficient pump, for he wrote to Cullen in August 1755 asking him to perform the experiment using a recently acquired pump at Glasgow:

> "I have endeavoured to get an experiment made here with the air pump, to no purpose. Dr Stewart's is not in order, and his operator is so slow and surly that I am quite tired of dunning him. Mr Wilson told me that he had sent one to Glasgow, which he was to set up immediately. If it be in order, I beg you will desire Mr torbet to try the following experiment, and to let me know the result of it as soon as he can conveniently do it."[18]

John Stewart was professor of natural philosophy at Edinburgh from 1742 to 1759. This implies that there existed no pump to which Black might have access in the chemistry laboratory; or if there was, then Andrew Plummer was not willing to allow him to use it.[19] Unless Stewart's pump later found its way into the chemistry laboratory there is no reason to associate the Playfair Collection pump with Black.

NOTES AND REFERENCES

1. J R Partington *A History of Chemistry 2* (London 1962) 515.
2. For the early history of the air pump, see: John Robison *A System of Mechanical Philosophy 3* (London 1822) 566; George Wilson *Edinburgh New Philosophical Journal 46* 330 (1849); E N da C Andrade *Proc Royal Institution 26* 114 (1929-31); Maurice Daumas *Scientific Instruments of the Seventeenth and Eighteenth Centuries and their Makers* (London 1972) 217.
3. Henry Guerlac 'Francis Hauksbee' in C C Gillispie (ed) *Dictionary of Scientific Biography 6* (New York 1972) 169. Hauksbee's first demonstration to the Royal Society involved the use of an air pump: this is recorded in the Society's manuscript Journal Book, Volume 4 (covering the period 1702-14), page 37.
4. Francis Hauksbee *Physico-Mechanical Experiments on Various subjects* (London 1709) 1.
5. Henry Guerlac 'Sir Isaac and the ingeneous Mr Hauksbee' in I Bernard Cohen and René Taton (eds) *Mélanges Alexandre Koyré: L'aventure de la Science 1* 241 (Paris 1964).
6. William Vream *Description and Use of Mr Hauksbee's Air Pumps* (London 1717) 7.

7. J T Desaguliers *A Course of Experimental Philosophy 2* (London 1744) 378. Desaguliers refers to Vream as "my Operator for Philosophical Machines."

8. Defining a 'Hauksbee-type pump' as a free-standing, double barrelled vacuum pump operating by a rack and pinion mechanism (although the later table air pumps functioned in a fundamentally similar way) I have been able to discover only four (or possibly five) examples surviving, locations as follows: Science Museum, London (on loan from the Royal Society, inv no 1970-24); Longleat House, Wiltshire (property of the Marquess of Bath); Museum of the History of Science, Oxford (see C R Hill *Museum of the History of Science Catalogue 1: Chemical Apparatus* (Oxford 1971) item 87); Deutsches Museum, Munich. In addition to these, Daumas claims that an example is to be found at the Academie des Sciences, Paris (*op cit*(2) 234). This instrument could not be traced in 1971, however.

9. Details of the rack and pinion mechanisms are as follows:

Pump	Teeth on pinion	Teeth on rack	Ratio
Playfair Collection	32	16	2
Museum of the History of Science	32	32	1
Science Museum	32	32	1
Deutsches Museum	32	32	1
Longleat House	24	32	2/3

10. Benjamin Martin *The Description and Use of a New, Portable, Table Air Pump* (London 1766) 1. Martin may have been referring to the large exhausting and compressing pumps of a type produced by George Adams, an example of which, constructed in 1762, survives in the King George III Collection at the Science Museum, London (inv no 1927-1624); see J A Chaldecott *Handbook of the King George III Collection of Scientific Instruments* (London 1951) 32 item 101. An earlier single barrelled variety is John Smeaton's pump of 1752 *ibid* 31, item 100.

11. Many examples of this type are to be found. See, for instance, G L'E Turner and T H Levere *Martinus van Marum... Volume IV, Van Marum's Scientific Instruments* (Leyden 1973) 230, item 139.

12. Cohen and Taton *op cit* (5) 238.

13. Probable dates on stylistic grounds have been suggested by Mr M C Baker, Department of Art and Archaeology, Royal Scottish Museum, private communication.

14. William Cullen *Essays and Observations, Physical and Literary 2* 145 (1756).

15. 'Inventory of Instruments for Philosophical Experiments And Observations belonging to the University of Glasgow, As they were delivered by Mr Lowdroun to Mr Dick Feb 13 1727'. Glasgow University Archives, MS 5291.

16. Glasgow University Library, Cullen papers, MS sheet headed 'Memorandum for Mr John Allan'.

17. Joseph Black *Essays and Observations, Physical and Literary 2* 157 (1756).

18. John Thomson *Account of the Life, Lectures and Writings of William Cullen M.D. 1* (Edinburgh 1832) 579. About Stewart it has been said "[he] has left behind him no trace of his acquirements or teaching", see Alexander Grant *The Story of the University of Edinburgh 1* (Edinburgh 1883) 272.

19. Plummer's apparatus was, according to Black, "very imperfect for a Course of Philosophic Chemistry", see Thomson *ibid* 93.

2. Pneumatic trough

[Pneumatic Trough used by Dr Joseph Black in the collection of fixed air (Carbonic Acid) and other gases. 1858.275.2]

Diameter 650, depth 105mm

Trough constructed of wooden staves bound by two riveted iron bands, and a wooden base (mahogany).

Figure 15 Pneumatic trough

The earliest forms of pneumatic trough[1] (through which, when filled with water, gases are bubbled to be collected in water filled receivers) were originally described by John Mayow in 1674[2] and Stephen Hales in 1727.[3] The latter used a wooden trough of barrel-like construction. Joseph Black did not describe such an apparatus in his investigations on carbon dioxide.[4] However, in Robison's edition of Black's lectures, there is a reference to the collection of carbon dioxide "in Dr Hales's manner" using a "tub of water"[5], a reference to the collection of hydrogen in "inverted glasses filled with water"[6] and a figure showing the collection of gases in a pneumatic trough.[7]

George Wilson, in 1859, considered this trough to have been used by Joseph Black.[8] Although it might be thought that the trough is rather shallow for its purpose, Samuel Parkes described a wooden example of similar depth in his *Chemical Catechism*.[9] Ceramic troughs became increasingly common in the 19th century.

Black referred to an object rather similar to the Playfair Collection trough when writing to James Watt in August 1786:

> "... I enquired of you whether the hard oil Vernish (Vernis Martin) which is put on Snuff Boxes and Tea-trays is sold at a reasonable price at London or Birmingham. I have a Mahogany Cistern or Tub 2 feet wide and a few inches deep which I want to vernish—..."[10]

The trough was exhibited at the Loan Collection of Scientific Apparatus at South Kensington in 1876[11] and at the International Exhibition in Glasgow in 1901.[12]

NOTES AND REFERENCES

1. John Parascandola and Aaron J Ihde 'History of the Pneumatic Trough' *Isis 60* 351 (1969); Lawrence Badash 'Joseph Priestley's Apparatus for Pneumatic Chemistry' *J Hist. Med. 19* 139 (1964).
2. John Mayow *Tractatus Quinque Medico-Physici* (Oxford 1674) 96, plate 5.
3. Stephen Hales *Vegetable Staticks* (London 1727) 183.
4. There is no description of a trough in either Black's MD thesis *De Humore acido a Cibis orto, et Magnesia Alba* (Edinburgh 1754) or his paper 'Experiments upon Magnesia Alba, Quicklime, and Some other Alcaline Substances' in *Experiments and Observations, Physical and Literary 2* 157 (1756). There is an account of an experiment performed by Black in 1757 or 1758 in which carbon dioxide was collected in a bladder (ie not over water in a trough), see Joseph Black *Lectures on the Elements of Chemistry ... Published by John Robison 2* (Edinburgh 1803) 112. Carbon dioxide is slightly soluble in water (a fact known to Black) though it can be collected over a pneumatic trough.

5. Joseph Black *Lectures on the Elements of Chemistry ... Published by John Robison 2* (Edinburgh 1803) 66.
6. *Ibid* 223.
7. *Ibid,* plate II, M. This is unlikely to be a representation of one of Black's troughs as Robison precedes the plates with the remark "A few figures being added to those expressly alluded to in the Lectures".
8. George Wilson suggested the association with Black in his annual report for 1858, published as Appendix F to the *Sixth Report of the Science and Art Department to the Committee of Council on Education* (London 1859). See also Wilson's address as President of the Royal Scottish Society of Arts, delivered on 23 November 1857 when the trough was still in the possession of William Gregory, *Transactions of the Royal Scottish Society of Arts 5* 53 (1861).
9. Samuel Parkes *Chemical Catechism* thirteenth edition (London 1834) 544.
10. E Robinson and D McKie (eds) *Partners in Science* (London 1970) 154, letter 108 (Black to Watt, ca August 1786).
11. *Catalogue of the Special Loan Collection of Scientific Apparatus at the South Kensington Museum* (London 1876), item 2537.
12. *Official Catalogue of the Scottish History and Archaelogy Section, International Exhibition Glasgow 1901* (Glasgow nd), item 770(a).

3. Balance

[Balance used in his experiments by Dr Joseph Black, who was the first to employ the balance in chemical investigation. 1858.275.3]

Overall height (to top of arm) 663, base 566 x 261, beam length 432 mm.

The beam (which has swan-neck beam ends), the pans and the suspension arm are of iron and the base on which these are mounted is of soft wood. The drawer contains five compartments one of which can accommodate the beam. A paper label is attached to the top of the base which bears the manuscript inscription "Balance used in his experiments by Dr Joseph Black". The pan cords are replacements (though the original ones survive) and these are gathered below the suspension with sealing wax and lead collars.

Figure 16 Balance with lead block (object 4) in drawer

Although the balance had been in use by assayers for many centuries before, it was during the 18th century that the chemical balance evolved to become a precision instrument. It has been said that it was Joseph Black who "first provided examples of reasoned sets of chemical experiments in which the balance was used at almost every stage".[1] The Playfair Collection balance was known as Black's balance when it came to the Industrial Museum of Scotland in 1858 and it remains the most renowned single item of the collection.

An early form of hydrostatic balance, from which chemical balances evolved, was described and illustrated in John Harris' *Lexicon Technicum* of 1710[2] (it seems certain that Andrew Plummer, professor of medicine and chemistry at Edinburgh from 1726 to 1755, once possessed the example which has passed down through his family[3]). Low accuracy in weighings was largely due to the swan-neck bearings from which the pans were hung, and the suspension and non-rigidity of the beam. It was not until bearings were replaced by steel or agate planes and knife edges and beams were constructed on new principles that accuracy could be increased.[4] Significant advances in design were made in balances constructed by Thomas Harrison for Henry Cavendish ca 1775, by Pierre Bernard Mégnié and Nicolas Fortin for Antoine Lavoisier ca 1785 onwards and by Jesse Ramsden for the Royal Society in 1789.[5] However, the Playfair Collection balance does not incorporate any refinements introduced in these balances and it may indicate an earlier date of construction. Lyon Playfair referred to it in 1876 as a "grocer sort of balance... no better than a pair of grocer's scales".[6] The balance is indeed of a type used by apothecaries and merchants throughout the 18th century (and later) with typical swan-neck beam ends. It is somewhat larger than most contemporary examples and is mounted with an iron support bar and pillar on a wooden base which contains a drawer with compartments (probably for storing weights), including one for housing the disassembled beam.

Lyon Playfair was somewhat uncertain about the balance's provenance when he referred to it at his inaugural lecture as professor of chemistry at Edinburgh in 1858:

> "A strange and clumsy balance it was, if it be the same or resemble that which was known to be in common use in his laboratory . . .
> This famous old balance, for which the traditional evidence is quite consecutive and satisfactory was among the lumber of the laboratory on my accession to office; but in future, will occupy a distinguished place in the National Museum, so worthily presided over by my colleague Dr G.Wilson."[7]

George Wilson, first Director of the Industrial Museum of Scotland, had no such doubts.[8] Objections to the provenance of the balance might be raised on the grounds that the balance was not sensitive enough for Black's research (though the survival of this balance does not preclude the ownership of others of greater sensitivity). In Black's classic work on magnesia alba[9] and in his later work on quick-lime[10], weights are given to the nearest grain (approximately 65 milligrams) while weighing, typically, 3 drachms (180 grains). Clearly accuracy greater than this was difficult to attain because Black comments "a light sediment . . . when collected with the greatest care, and dryed, weighed, as nearly as I could guess, one third of a grain".[11] 'Black's balance' would have been adequate for this work: a recent investigation of its present state[12] concluded that an out-of-balance of about one grain can be detected when a load of 200 grains is added to each of the pans. It was further pointed out that the poor rest point reproducibility can be accounted for by wear and corrosion of the knife and its bearings; a balance in this state would normally have been sent for knife resharpening and bearing rereaming.

In Black's work on latent heat, weighings are given to less accuracy. For instance in a lecture by Black on the latent heat of ice from 1762 onwards, accuracy to only the nearest half drachm is given.[13] However, in later work on the analysis of water samples the Playfair Collection balance would not be nearly sensitive enough. In the analysis of the waters of some hot springs in Iceland, Black gives weighings of sediments to the nearest 0.01 grain (in 0.4 grains) and to 0.1 grain (in 38 grains).[14] For the former weighings it is highly likely that he was using a small beam balance with a rider which he had invented and described in a letter to James Macie in 1790: Black claimed that the balance could read from one grain down to $\frac{1}{1200}$ grain.[15] It is not clear how Black could have undertaken the latter weighings: certainly both his "very sensible" beam balance and the Playfair Collection balance would have been unsuitable. Jesse Ramsden's balance had already been constructed by this date but there is no evidence that a costly instrument of this type had been acquired by Black.[16]

Summing up, the Playfair Collection balance may well have been used by Black in some of his experiments and in his lecture demonstrations, but other balances must have been available if his experimental weighings are to be accounted for. Towards the end of Black's career it must have seemed very clumsy and outdated as a chemical balance.

The balance was exhibited at the Special Loan Collection at South Kensington in 1876[17], at the International Exhibition in Glasgow in 1901[18] and the 'Edinburgh and Medicine' exhibition in Edinburgh in 1976.[19] Copies of the balance exist at the Science Museum, London[20] and at the National Museum of History and Technology, Smithsonian Institution, Washington. A reconstruction of Black's beam balance has been made for the Royal Scottish Museum, Edinburgh.[21]

NOTES AND REFERENCES

1. John T Stock *Development of the Chemical Balance* (London 1969) 2.
2. John Harris *Lexicon Technicum 2* (London 1710). The balance is referred to as "an Hydrostaticall-Ballance, for finding the Specifick Gravities of Liquids and Solids with ease and accuracey. By F Hauksbee, in Vine-Office-Court in Fleet-Street".
3. R G W Anderson and A D C Simpson *Edinburgh and Medicine* (Edinburgh 1976) 35, item 130. See Appendix 3, item 1, of this work.
4. Maurice Daumas *Scientific Instruments of the Seventeenth and Eighteenth Centuries and their Makers* (London 1972) 221. A rigid beam constructed of a hollow double cone was used by Ramsden. A very early one built on a framework principle was described by J Hyacinthe Magellan 'Lettre sur les Balances d'Essai' *Obs sur la Physique 17* 43 (1781)

5. The balance associated with Henry Cavendish is at the Royal Institution, London (copy at the Science Museum, inv no. 1930-770); several of Lavoisier's balances are at the Conservatoire National des Arts et Metiers, Paris (there is also a balance by Fortin at the Whipple Museum of the History of Science, Cambridge, acc no. 2230); Ramsden's balance is at the Science Museum, inv no. 1900-166.

6. Lyon Playfair 'On Air and Airs' *Free Evening Lectures delivered in connection with the Special Loan Collection of Scientific Apparatus, 1876* (London nd) 144.

7. Lyon Playfair *A Century of Chemistry in the University of Edinburgh* (Edinburgh 1858).

8. Appendix F to the *Sixth Report of the Science and Art Department of the Committee of Council on Education* (London 1859).

9. Joseph Black *De Humore Acido . . . et Magnesia Alba* (Edinburgh 1754).

10. Joseph Black 'Experiments upon Magnesia Alba, Quicklime, and other Alcaline Substances' *Essays and Observations, Physical and Literary 2* 157 (1756).

11. *Ibid* 196.

12. Report by John T Stock, University of Connecticut, dated 16 January 1969.

13. Joseph Black *Lectures on the Elements of Chemistry . . . published by John Robison 1* (Edinburgh 1803) 124.

14. Joseph Black 'An Analysis of the Waters of some Hot Springs in Iceland' *Trans. Royal Soc. Edinburgh 3* 95 (1794).

15. Draft of a letter from Joseph Black to James Louis (sic) Macie dated 18 September 1790, Edinburgh University Library MS 873 volume 3, f 158; published in *Annals of Philosophy* (New Series) *10* 52 (1825); described in S F Gray *The Operative Chemist* (London 1828) 234 and M Faraday *Chemical Manipulation* 3rd ed (London 1842) 62. The recipient of the letter, James Lewis Macie later changed his name to James Smithson and endowed what is now the Smithsonian Institution, Washington. The letter is also reprinted in W J Rhees (ed) *The Scientific Writings of James Smithson Smithsonian Miscellaneous Collections* (Washington 1879) 117. It is likely that the balance was developed specially for this investigation, for in the letter Black writes "I am employed in examining the Iceland waters, but have been often interrupted". Black's paper was read to the Royal Society of Edinburgh ten months after the letter was written, on 4 July 1791. When Black analysed the Water of Leith in July 1784 (Edinburgh University Library MS Gen 873 volume 2, f 182) he gave weighings only to the nearest ¼ grain and so presumably had not developed his beam balance by this date. Black's beam balance is discussed by Max Speter in *Zeitschrift für Instrumentenkunde 50* 204 (1930).

16. Ramsden's balance (see ref 5) was accurate to the nearest 0.01 grain when loaded to ten pounds. (0.6 milligrams in 4536 grams).

17. *Catalogue of the Special Loan Collection of Scientific Apparatus at the South Kensington Museum* (London 1876), item 2401.

18. *Official Catalogue of the Scottish History and Archaeology Section, International Exhibition Glasgow 1901* (Glasgow nd), item 770b.

19. Anderson and Simpson *op cit* (3) 40, item 185.

20. A Barclay *Catalogue of the Collection in the Science Museum . . . Chemistry* (London 1927) 15, plate I.

21. Anderson and Simpson *op cit* (3) 61, item 466 (reg no 1971-272).

4. Lead block

[Counterpoise (lead weight) for snow pan & lid (?) suspended from arm of Dr Blacks balance wanting one of the scales. 1858.275.4]

Dimensions 45 x 38 x 38 mm.; weight 653 gms.

Cuboidal mass of lead.

The lead weight has been associated with 'Black's balance', though its function is not clear. The 'snow pan and lid' with which it is associated in the register entry does not form part of the Playfair Collection. The term is not one used elsewhere. It is possible that it was some type of calorimeter and that it was used in a lecture demonstration to determine the latent heat of ice. Presumably one of the scale pans of the balance would be removed and would be replaced by the vessel which would be suspended from the beam. The counterpoise would bring the balance back into equilibrium and the addition of ice and water could be directly weighed. Black described an experiment of this nature (though he did not give details of his apparatus) and he mentioned that he had explained it to his classes in Glasgow and Edinburgh in every session from 1762.[1]

Another possible use of the weight might be its use as a counterpoise in hydrostatic experiments. Most sets of balances incorporated counterpoises which balanced a pear-shaped glass weight which was then immersed in the liquid being examined.[2] An argument against this possibility is that the weight is considerably heavier than those in other remaining cases of hydrostatic balances.

The register entry is so specific about the function of the weight that the experiment is possibly one which Lyon Playfair or George Wilson recalled being performed.

NOTES AND REFERENCES

1. Joseph Black *Lectures on the Elements of Chemistry . . . Published by John Robison 1* (Edinburgh 1803) 124.
2. See, for example, J T Desaguliers *A Course of Experimental Philosophy 2* (London 1744) 194, plate 18.

5. Double bellows
[Blow-pipe bellows invented by Sir Humphrey (sic) Davy. 1858.275.5]

Base 381 x 273mm.

A wooden (mahogany) base supports two bellow chambers hinged on opposite sides; the bellows material is leather. An iron counterweight is attached to one chamber by an adjustable grooved iron arm fixed with a brass screw. The other chamber holds an iron arm for attaching to a pedal (missing) by means of a cord.

Figure 17 Double bellows

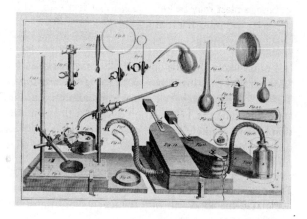

Figure 18 Portable laboratory, from A F Cronstedt
An Essay towards a System of Mineralogy... improved by John Hyacynth de Magellan 2 **(London 1788) plate 2**

The blowpipe was developed for use in the chemical analysis of minerals by Swedish chemists in the mid-18th century; the first paper on the subject was published by Sven Rinman in 1746.[1] Joseph Black referred to the blow-pipe as "an indispensible instrument for every mineralogist, and various contrivances have been introduced for accommodating it for an imitation of almost every metallurgical operation."[2] Black referred to the instrument as being used for small-scale glass blowing operations, sometimes in conjunction with bellows: "Those who make thermometers with it, often blow it with bellows, but the mouth is most ready and most under command."[3]

The use of bellows to provide a stream of air for the blowpipe instead of using the lungs was suggested by Carl Friedrich Zimmermann in 1739.[4] However, bellows with a single blowing chamber would give an intermittent blast of air. Progression to a double bellows seems to have been first suggested in 1788 in an appendix by J H de Magellan[5] to an English edition of A F Cronstedt's *System of Mineralogy*:

"the operator blows through the mouth-piece of the blow-pipe . . . But if the operator has the double bellows . . . he fixes them to the table by the brass clamp . . .

If he works with his foot on the pedal, the string of which is seen hanging from the end of the bellows. . . (and is always up on account of the weight) then the air is absorbed by the bellows, from whence it is propelled by the motion of the foot on the pedal to the [other] bellow, whose constant weight drives it out. . .

N.B.1 This double bellows is packed up by itself in a mahogany case, about 9 inches long, 6½ wide, and about 3¼ deep, outside measure."[6]

The plate which accompanies it (fig 18) shows it to be very similar to the Playfair Collection example. The dimensions are, however, smaller. The apparatus described was sold "at moderate prices, by William Brown, Bookseller, at the Corner of Essex-street, in the Strand, near Temple-Bar, London".[7]

A better description of the double bellows appears in Frederick Accum's *Descriptive Catalogue* of 1817, where it was offered complete with blowpipe for six guineas:

> "The common practice of blowing with the mouth, by means of the blowpipe, though very ready, and requiring an instrument of inconsiderable cost, is not so easy in practice for everyone, nor so advantageous in effect as the extrusion of air by means of bellows with double partitions. Besides, the air respired from the lungs being loaded with moisture and contaminated with carbonic acid gas, is very unfit for maintaining combustion; and the act of blowing, even to the most skilful, is attended with fatigue. It requires a confinement of the hands and an awkward position of the head, which considerably diminishes the power as well as the ease of the operator. These disadvantages are obviated by the application of the blowpipe with double bellows. The apparatus, when firmly fixed to a table by means of a clamp, can be worked with the foot placed on the string attached to the pedal of the instrument. The air being then propelled through the pipe, a blast may be kept up with perfect ease, whilst the operator who sits before the lamp, has both his hands and mouth left at liberty."[8]

It is quite likely that the double bellows was acquired by Joseph Black. He mentioned, in his lectures, that "Engestroem describes a box, containing the blowpipe and other implements, which he calls the POCKET LABORATORY."[9] The University Library at Edinburgh acquired all three English editions of Cronstedt's *System of Mineralogy*, including the 1788 edition of de Magellan. An 18th or early 19th century date of origin for the Playfair Collection bellows is likely because a later and simpler form with the chambers mounted on top of one another instead of side-by-side, was described in 1803,[10] and this was adopted as the preferred design.

Thomas Charles Hope referred to double bellows in several places in the notes he used for his lectures, but always associated the apparatus with the name 'Nooth', for instance:

> "Various Machines contrived for furnishing constant stream of Air to Blowpipe-Double Bellows improved by Dr Nooth".[11]

Charles Hopson illustrated the bellows in his *General System of Chemistry* of 1789,[12] referring to them as 'Double Bellows invented by Dr Nooth', the figure being taken from de Magellan's work of a year earlier.

The bellows cannot have been invented by Humphry Davy as is suggested in the Museum Register entry; he was only ten years old when the apparatus was first described.

NOTES AND REFERENCES

1. Sven Rinman 'Anmarkning om en art Jern-haltig Tennmalm ifran Dannemora Sokn i Upland' *Kongl. Svenska Vetenskaps Academiens Handlingar* 7 176 (1746); see J R Partington *A History of Chemistry 3* (London 1962) 174, 175, 185.

2. Joseph Black *Lectures on the Elements of Chemistry... published by John Robison* 1 (Edinburgh 1803) 326. Black himself probably designed the variety of blowpipe which still bears his name: it is of conical form to provide a relatively large reservoir of air under pressure which gives a steady flame. J J Griffin, the Glasgow instrument maker, disputed such a claim: "The blow-pipe commonly called Dr Black's was the invention of a German workman living in Glasgow... the instrument which Dr Black used was made by Mr Crichton, the thermometer maker, to whom Dr Black took the foreigner's blowpipe for a pattern." (see J J Griffin *Chemical Recreations* 8th ed (Glasgow 1838) 111). However, James Crichton is not recorded as an instrument maker until 1785 (see D J Bryden *Scottish Scientific Instrument-Makers* (Edinburgh 1972) 46) and the conical form of the blowpipe is clearly shown in a sketch made by T C Hope while attending Black's lectures in 1782 (Edinburgh University Library MS Dc.10.92).

3. Joseph Black *op. cit.* (2) 326. Black used a 'pair of chamber-bellows' in an experiment (performed in 1756) to prepare carbon dioxide by blowing air over red hot charcoal (*ibid 1* 88).

4. J R Partington *A History of Chemistry 2* (London 1961) 711.

5. There is no evidence that J H Magellan (1722-1790) invented the bellows. He can be most kindly described as a communicator of contemporary scientific knowledge, though possibly more accurately as a plagiarist. See *DNB* and J R Partington *A History of Chemistry 3* (London 1962) 248.

6. A F Cronstedt *An Essay towards a System of Mineralogy... Translated by Gustav von Engestrom... Second Edition, greatly enlarged and improved by John Hyacinth de Magellan* 2 (London 1788) 994.

   The first and second editions of the English translations of Cronstedt's *System of Mineralogy* (1770 and 1772) both contain descriptions of the 'pocket laboratory', a portable kit of chemicals, blowpipe etc developed for conducting chemical analyses while involved in fieldwork. However the double bellows was not described until de Magellan's edition of 1788 (see

W A Smeaton 'The Portable Chemical Laboratories of Guyton de Morveau, Cronstedt and Gottling' *Ambix* 13 84 (1965)). A fairly complete kit (including double bellows) which fits into a single box 12½ x 8 x 6¼ inches survives at Transylvania University, Lexington, Kentucky, see Leland A Brown *Early Philosophical Apparatus at Transylvania College* (Lexington 1959) 22.

7. *Ibid* 1015.

8. The catalogue is printed at the end of Frederick Accum *Chemical Amusement* (London 1817).

9. Joseph Black *op cit* (2) 326.

10. C R Aikin 'Description of a Portable Chamber Blast Furnace' *Philosophical Magazine* 17 166 [1803]; William Henry *An Epitome of Chemistry* 4th ed (London 1806) *Appendix 1* vi, plate VI.

11. Edinburgh University Library, MS Gen 272 (found in notes amongst loose papers in this box). 'Nooth' is presumably John Mervin Nooth (MD Edinburgh 1766, d 1828). De Magellan made no acknowledgement to the inventor of the bellows (ref 6).

12. C R Hopson *A General System of Chemistry* (London 1789), plate II, figs 14 and 15.

6. Wire cage
[Cage for placing ice in the focus of a Parabolic Mirror. 1858.275.6]

Overall height 406, diameter of cage 172 mm.

Cage of iron wire supported by an iron rod which is screwed into a saucer-shaped iron dish mounted on a tripod. A circular hinged iron door above the cage is pierced with seven holes.

Figure 19 Wire cage

This is one of a number of pieces of apparatus developed from the 17th century onwards for demonstrating the properties of radiant heat. It was used in conjunction with parabolic mirrors, air thermometers and differential thermometers (represented in the Playfair Collection by objects 9-11 and 28-30).

The first experiment of the kind to discover whether 'cold' could be reflected from mirrors was probably performed by the Accademia del Cimento of Florence in 1660 although the use of a lens to focus 'cold' may have been made half a century earlier.[1] The Accademia described their experiment as follows:

> "The desire came to us to find out by experiment whether a concave mirror exposed to a mass of 500 pounds of ice would make any perceptible reverberations of cold in a very sensitive 400-degree therometer located at its focus. In truth this at once began to descend, but because the ice was nearby, it remained doubtful whether direct cold or that which was reflected cooled it most. The latter was taken away by covering the mirror, and (whatever the cause may have been) it is certain that the spirit immediately began to rise again".[2]

A related experiment was performed by M A Pictet in 1790:

> "I disposed the apparatus exactly as in the experiment for the reflection of heat, and employed the two mirrors of tin, at the distance of 10½ feet from each other. At the focus of one was a thermometer of air, which was observed with the necessary precautions, and at the focus of the other a matrass full of snow. At the instant the matrass was placed for experiment, the thermometer at the opposite focus descended several degrees, and remounted as soon as the matrass was removed".[3]

In this experiment, Pictet refers to a matrass (a glass vessel with a long neck) for containing the snow. However in a similar experiment using heated bodies, he refers to an apparatus similar to the Playfair Collection cage:

> "The bullet, of which we have already spoken, heated to a certain degree, but not sufficiently to become luminous, is then placed in the basket of iron wire at the focus of the first mirror".[4]

The experiment was performed in Edinburgh by the professor of natural philosophy, John Leslie (1766-1832) who described his cold source as "hollow cubes of block tin, formed exactly and planished".[5] He went on to complain that "experiments with Cold, though perfectly consonant, are much more troublesome in the execution, and require greater attention and stricter observation."[6]

There is also a record of the experiment being performed by Thomas Charles Hope for Count Rumford (Benjamin Thompson) in 1800, though on this occasion a glass bulb was used to contain the cold source:

> "I found myself in the company of Professor Hope (the successor of the celebrated Black), Professors Playfair and Stewart, and several other persons. We repeated the experiment which Pictet undertook ... Two metallic mirrors fifteen inches in diameter, with a focal length of fifteen inches, were placed opposite each other, sixteen feet apart. When a cold body (for example a glass bulb filled with water and pounded ice) as was the case on this occasion, was placed in the focus of one of the mirrors, and a very sensitive air thermometer was placed in the focus of the other mirror, the latter thermometer began immediately to fall."[7]

In the manuscript notes he used for lecturing, Hope makes reference to a "Globular Cage" with which he demonstrated the focusing of heat in his lecture on 'Incandescence':

> "I employ Charcoal in this Globular Cage. Its incandescence may be maintained & Invigorated by the bellows — ... I employ two concave metallic Mirrors or Speculums — I place them exactly opposite at the distance of 20 feet — and place the Cage in the focus of one, & the Object to prove the Heat in the focus of the other."[8]

A wire basket on a stand, used for containing hot bodies in the same experiment, acquired by Martinus van Marum, survives at Teyler's Museum, Haarlem, probably dating from 1825.[9] Black did not describe the experiment and it is likely that the cage in the Playfair Collection was acquired by Hope. The experiment using a red hot sphere contained in a wire cage was described in an edition of *Ganot's Physics* as late as 1890.[10] It seems more likely that the Playfair Collection cage was used for this demonstration rather than for containing ice in the 'radiation of cold'.

NOTES AND REFERENCES

1. W E Knowles Middleton *The Experimenters: A Study of the Accademia del Cimento* (Baltimore 1971) 205, note 222. The experiment with a lens can be dated to pre-1611, according to Middleton.
2. *Ibid* 205. An artist's impression of this experiment is reproduced in Philip D Thomson and Robert O'Brien *Weather* (1970) 118.
3. M A Pictet *An Essay on Fire* (London 1791) 86 (translation of M A Pictet *Essais de Physique 1* (Geneva 1790)).
4. *Ibid* 23.
5. John Leslie *An Experimental Inquiry into the Nature and Propogation of Heat* (London 1804) 4.
6. *Ibid* 23.
7. Sanborn C Brown (ed) *The Collected Works of Count Rumford 1* (Cambridge, Mass 1968) 477.
8. Edinburgh University Library MS Gen 268, envelope 38 ('Incandescence 1st'). The cage is also mentioned in a manuscript list of apparatus compiled by Hope as instructions to his assistant in preparing for the lecture (MS Gen 270, envelope 116 ('Lists of Specimens etc ')).
9. G L'E Turner and T H Levere *Martinus van Marum ... Volume IV. Van Marum's Scientific Instruments* (Leyden 1973) 266, item 214.
10. A Ganot *Elementary Treatise on Physics ... Translated by E Atkinson* 13th ed (London 1890) 391.

7. Pyrometer
[Pyrometer Ferguson's. 1858.275.7]

Overall height 356, base 1155 x 152, length of metal rod 888 mm.

A wooden (mahogany) base supports a brass stand on which a copper rod rests, one (flat) end touching a brass adjustment screw and the other (chisel-shaped) end resting against a lever mechanism which pushes a pointer over a silvered brass scale which is graduated 0 to 90 in single units. In operation the rod is heated by gas burnt at orifices along a tubular burner beneath the rod. An ivory maker's label is inset in the base and is inscribed "J. DUNN Edinburgh".

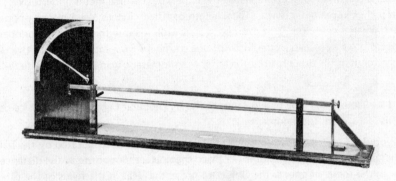

Figure 20   Pyrometer by J Dunn, Edinburgh

Quantitative experiments on the expansion of metals were carried out first in the early 18th century; among the early workers were George Graham and John Harrison. Petrus van Musschenbroek introduced the name 'pyrometer' in 1731 in a paper describing results of experiments designed to compare thermal expansions of different metals by heating rods under similar conditions.[1] In 1735 Cromwell Mortimer attempted to use the property of thermal expansion of metals to measure temperatures.[2]

Apart from Josiah Wedgwood's instrument of 1781, which used the property of the contraction of ceramic pieces to measure the temperature of furnaces[3], pyrometers of this period were of little value as temperature measurers. The pyrometer developed as a demonstration instrument to compare the expansion of different metals and these forms should be referred to, more accurately, as dilatometers. Many different types were produced from the mid-18th to the mid-19th centuries.[4]

The Playfair Collection example is an instrument designed simply for the demonstration of the variation of coefficients of thermal expansion of different metals (though only one bar of metal remains). The mechanism of operation shows similarities to Samuel Frotheringham's metalline thermometer of 1747[5] and an example illustrated in *Ganot's Physics*.[6] The latter functioned by means of a spirit lamp, whereas the Playfair pyrometer was heated by a series of flames from gas jets. The scale is very large, and may well have been commissioned by Thomas Charles Hope to be visible to a large audience.[7] Hope discussed pyrometers in his lectures:

> "For higher heats, substances which can sustain them are of course resorted to & metals alone are employed. Such instruments are called Pyrometers i.e. measures of fire; but though Mushenbrock-Ellicot-Smeaton, De Luc, Ramsden [,] Morveau [and] Daniel have suggested different forms, no contrivance nearly approaches the Thermr. in convenience or perfection."[8]

The pyrometer is the only signed instrument in the Playfair Collection. Its maker, John Dunn, was a good instrument maker whose business is recorded at a number of Edinburgh addresses over the period 1824 to 1842.[9] He attended the Edinburgh School of Arts[10] for whom he later constructed an air pump in July 1826. He made another pump for the chemical class at University College, London in December 1827[11] and was appointed Curator of the Museum of the Society of Arts (later the Royal Scottish Society of Arts) in 1833, holding the post until his death in (?) 1841.[12] Dunn was patronised by William Gregory (1803-1858) when the latter was setting up as a private lecturer in chemistry in Edinburgh but there is no record in his account book that he bought a pyrometer from Dunn.[13] When Gregory returned from Aberdeen in 1844 as professor of chemistry, the firm had ceased to exist.

The operation of the Playfair Collection pyrometer requires a supply of gas for heating the rod under investigation. The first mention of a supply laid on for the College appears to be in 1829[14] though from 1820 until its failure in 1827, the Edinburgh Portable Gas Company provided compressed gas derived from oil in portable urns[15] for which piping was unnecessary. It is likely that early piped supplies of gas were for lighting only: the first reference to a supply of gas laid on specifically for the chemistry classroom does not occur until 1858[16] and it was not until this time that gas burners became commonly used in chemical laboratories following Justus Liebig's development of an efficient burner in 1855.[17] The pyrometer cannot be dated any more accurately than having been made between the dates known for John Dunn's scientific instrument making activities, 1824 to 1841.

NOTES AND REFERENCES

1. Petrus van Musschenbroek *Tentamina Experimentorum Naturalium Captorum in Academia del Cimento* (Leyden 1731) 12.

2. J A Chaldecott 'Cromwell Mortimer, FRS (c1698-1752) and the Invention of the Metalline Thermometer for measuring High Temperatures' *Notes and Records of the Royal Society 24* 113 (1969).

3. J A Chaldecott 'Josiah Wedgwood (1730-95) – Scientist' *British J Hist Sci 8* 1 (1975).

4. Examples survive at the Science Museum, London (J A Chaldecott *Heat and Cold Part II Descriptive Catalogue* (London 1954), The Boerhaave Museum, Leyden (C A Crommelin *Descriptive Catalogue of the Physical Instruments of the 18th Century* (Leyden 1951) 45, and Teyler's Museum, Haarlem (G L'E Turner and T H Levere *Martinus van Marum . . . Volume IV Van Marum's Scientific Instruments* (Leyden 1973) 267).

5. Chaldecott *op cit* (2) 122.

6. A Ganot *Elementary Treatise on Physics . . . Translated by E Atkinson* 10th revised ed (London 1881) 251.

7. Alexander Grant *The Story of the University of Edinburgh 2* (Edinburgh 1884) 398. ("To be visible to a class of 500 Students, his [Hope's] experiments required to be performed on a very large scale.")

8. Taken from notes which Hope used for lecturing, Edinburgh University Library MS Gen 269 envelope 44 ('Expansion').

9. D J Bryden *Scottish Scientific Instrument Makers 1600-1900* (Edinburgh 1972) 48. It is known that Dunn had died by 31 October 1841 (see ref 12); the compiler of the 1842 Edinburgh Directory did not make the necessary deletion for that year.

10. *Glasgow Mechanics Magazine 5* 180 (1826-27).

11. John Dunn 'Description of an Improved Air-Pump' *Edin New Phil J 4* 382 (1828). The pump was presumably ordered by Edward Turner, who had moved from Edinburgh to London in 1827 to become first professor of chemistry at University College, see chapter 4.

12. 'List of Fellows . . . of the Royal Scottish Society of Arts', Appendix (page 9) to *Transactions of the Royal Scottish Society of Arts 1* (1841).

13. Aberdeen University Library MS 2206.24 'Account for Expenses for Apparatus &c incurred in setting up as a Teacher of Chemistry by Dr William Gregory beginning Martinmas 1829 Edinburgh.'

14. City of Edinburgh District Council Archives, MS Town Council Record volume 205, f104 (1829).

15. The demise of the Company is recorded in Edinburgh Central Library (George IV Bridge) MS YTP 733P (acc no G 81971) 'The Minute Book of the Edinburgh Portable Gas Company' (1827).

16. City of Edinburgh District Council Archives, MS Town Council Record volume 276, f51 (19 October 1858). Lyon Playfair made the request for improved laboratory facilities following his appointment.

17. Georg Lockemann 'The Centenary of the Bunsen Burner' *J Chem Ed 33* 20 (1956).

8. Model of a bimetallic spiral thermometer
[Thermometer formed of a spiral of brass. 1858.275.8]

Overall height 298, diameter of base 178mm.

A brass spiral strip is supported at its upper end by a vertical brass arm. An iron pointer attached to the lower end lies above a brass annulus which rests on three spherical feet.

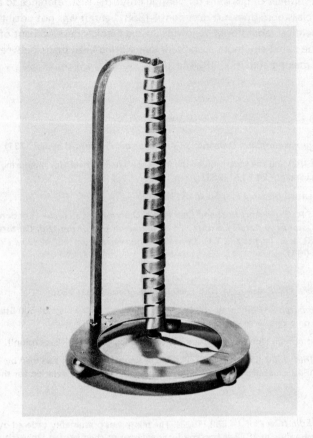

Figure 21   Model of a bimetallic spiral thermometer

Thermometers which functioned by the distortion of a metallic strip composed of two metals with different coefficients of expansion[1] were first described by J H Lambert in a publication of 1779.[2] The instrument maker James Crichton of Glasgow described a bimetallic thermometer in 1803[3] and Thomas Charles Hope possessed one of his though it did not form part of the Playfair Collection and no longer survives.[4] In some bimetallic thermometers the strip was coiled into a spiral which allowed for a longer strip to be used, increasing the effect. The first description of such a thermometer appears to have been made by J H de Magellan in 1782:

> "To construct it, a strip of steel and one of brass are solidly riveted or even soldered together; for heat and cold make this double strip curve more or less, according to the reading of the thermometer. It is better to give this strip a spiral form, so as to have it long enough to enlarge the movement. . . This strip, being fastened at one end, moves the pencil that is attached to the other end (or to a rod fastened to this) according to the different temperature of the atmosphere."[5]

Abraham Louis Breguet constructed the best known form of this type of thermometer, utilising a trimetallic strip of platinum, gold and silver from 1817, though he had considered using a bimetallic strip from 1802.[6] Hope discussed this latter type of thermometer in his lectures, though there is no evidence that he possessed one:

> "A Thermometer of extreme sensibility has been more lately manufactured at Paris, upon the same principle by Breguet — the two Bars one silver t. other Platinum are made very slender & are twisted into a

spiral form — The Upper Extremity is firmly fixed to a projecting portion of a frame & to the lower is attached an index, which ranges round the scale which is disposed in a circular form—

This instrument is indebted for its extreme sensibility to the length & Slenderness of the Bars—

It is, like the preceding, only adapted to Aerial Temperatures."[7]

The object in the Playfair Collection has the basic appearance of a spiral bimetallic thermometer though it could not function as one as the metal spiral is constructed from a single metal, brass. It is crudely made and its substantial spiral (much more massive than in authentic Breguet-type examples) indicates that it may have been intended for showing to a large class. It is quite likely that Hope had it specially made, for an *aide-memoire* amongst his lecture notes reads "Apparatus to be got this Summer — 1827 . . . Model of Breguets Thermr. Ord^d".[8] The instrument maker is not revealed, though the name of John Dunn is mentioned two items lower in this list (see object 7). An alternative possibility is that the instrument was made to function as a simple pyrometer, the length of metal being maximised by forming a spiral. However expansion of the brass strip alone is inadequate to provide visible rotation of the pointer (this has been tried over a large temperature range) and in any case the instrument is not graduated which further suggests its non-functionality.

NOTES AND REFERENCES

1. For a short history of the bimetallic thermometer see W E Knowles Middleton *A History of the Thermometer* (Baltimore 1966) 169.
2. J H Lambert *Pyrometric oder vom Maase des Feuers und der Warme* (Berlin 1779) 124. It is possible that an instrument was constructed by David Rittenhouse by 1767; see Knowles Middleton *op cit* (1) 170.
3. James Crichton 'On the Freezing Point of Tin. . . with a Description of a Self-registering Thermometer' *Philosophical Magazine* **15** 147 (1803).
4. Edinburgh University Library MS Gen 269 'List of Specimens'. This list, which includes a mention of Crichton's thermometer, was compiled for the benefit of Hope's assistant who assembled the apparatus for each day's lecture.
5. J H de Magellan *Obs. sur la Phys.* **19** 353 (1782).
6. Knowles Middleton *op cit* (1) 171.
7. Edinburgh University Library MS Gen 269, envelope 44 (notes from which Hope lectured).
8. Edinburgh University Library MS Gen 272.

9. Air thermoscope
[Air Thermometer. 1858.275.9]

Length of thermometer tube 286, diameter of bulb 58 mm.

A gilded glass sphere blown on the end of a glass tube protrudes through a cork held in the neck of a glass bottle which contains lead shot.

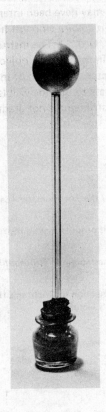

Figure 22   Air thermoscope

Air thermoscopes and thermometers show changes in temperatures by the expansion and contraction of air in a bulb causing the fall and rise of a liquid in the attached capillary tube. They were developed from the beginning of the 17th century, and the priority for their discovery has been discussed at some length.[1] These problems were appreciated by Joseph Black in his lectures, and he felt (as is generally now agreed) that the most probable inventor was Sanctorius (1561-1636):

> "The history of this invention [the thermometer] is a little obscure, and the contrivance of the first thermometer has been attributed to three or four different persons. Sanctorio, who is distinguished by the discovery of what is called insensible perspiration, appears, however, to have the best title to it."[2]

Although it is not clear whether Black demonstrated the air thermometer, Thomas Charles Hope noted the operation of the instrument while attending Black's lectures in 1782:

> "He [Sanctorio] took a hollow ball with a long tube open at the extremity — ... He plunged the end of the tube in to an open Vessel with some colored liquor as Spt of Wine"[3]

Later when he himself delivered the chemistry lectures, Hope had an example to display. He mentions in the notes from which he lectured "*I need* Scarcely have recourse to this Air Thermr to Show it".[4]

It is not clear why the bulb is gilded. John Leslie, professor of natural philosophy at Edinburgh, mentions gilding or enamelling one of the bulbs of a differential thermometer in 1804.[5] It may possibly have been used in association with an air thermometer with a plain glass bulb to demonstrate the effect of different surfaces to radiant heat.

NOTES AND REFERENCES

1. W E Knowles Middleton *A History of the Thermometer* (Baltimore 1966) 3.
2. Joseph Black *Lectures on the Elements of Chemistry . . . Published by John Robison* 1 (Edinburgh 1803) 50. In a footnote to this entry, Robison questioned Black's opinion.
3. Edinburgh University Library MS Dc.10.9.
4. Edinburgh University Library MS Gen 268, envelope 32.
5. John Leslie *An Experimental Enquiry into the Nature and Propagation of Heat* (Edinburgh 1804) 561.

10. Differential thermoscope ('pyroscope')
[Differential Thermometers or Thermoscopes (3 specimens). 1858.275.10]

Overall height 362, diameter of clear bulb 51, diameter of gilded bulb 57 mm.

Two glass spheres, one gilded, one clear, are blown on each end of a glass capillary tube bent with two right angles. This is supported in a groove cut in a square wooden base and fastened by two leather pieces. The instrument contains a dark fluid (coloured sulphuric acid?).

11. Differential thermoscope ('photoscope')
[Differential Thermometers or Thermoscopes (3 specimens). 1858.275.10]

Overall height 387, diameter of bulbs 32 mm.

Two glass spheres, one coloured green, the other clear, are blown on each end of a glass capillary tube bent through two right angles. This is supported in a groove in a wooden rod attached to a turned wooden base. The instrument contains a dark fluid (coloured sulphuric acid?).

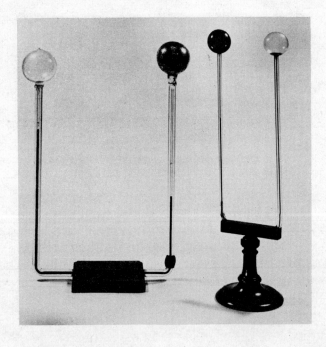

Figure 23 Differential thermoscopes: pyroscope (right), photoscope (left)

88

The differential air thermoscope or thermometer indicates or measures differences in temperature caused by radiant heat falling on two bulbs of air. The different heating effect causes one bulb of air to expand more than the other, causing the movement of a column of liquid in the capillary tube joining them. The instrument was first produced in the latter half of the 17th century.[1]

Thomas Charles Hope's colleague at Edinburgh, John Leslie, professor of natural philosophy and later of mathematics, claimed in a publication of 1813 that the first description of the instrument had been by him in 1804.[2] Indeed, Leslie was involved in 1812 in a controversy over the priority of invention of the differential thermometer, and though he appeared to have lost his case, it did not deter him from repeating the claim a year later.[3]

Leslie's first description of the instrument is as follows:

"Two glass tubes of unequal lengths, each terminating in a hollow ball and having their bores somewhat widened at the other ends, a small portion of sulphuric acid tinged with carmine being introduced into the ball of the longer tube, are joined together by the flame of a blow pipe, and afterwards bent into the shape of the letter U, the one flexure being made just below the joining... The balls are blown as equal as the eye can judge, and from four tenths to seven tenths of an inch in diameter".[4]

Leslie suggested several minor adaptations to his differential thermometer to measure different effects. One of these was his 'photometer'[5] which he first thought of in 1797. One of the bulbs was coated with china ink, and it was claimed that the instrument would "indicate the true calorific power of the incident rays [of light]".

In another modification, the instrument could be used for:

"... estimating with nice precision the intensity of the diffuse radiations of heat ... the ball of the graduated stem is completely gilt or enamelled with gold. But the two balls, exposed to the same influence, will now receive very different impressions, and the excess of energy, which the instrument marks, must therefore amount to nearly seven-eights of the whole vibratory tide. Hence it will measure the quantities of heat that are continually thrown from the fire into a room."[6]

This instrument was later termed a 'pyroscope' by Leslie.[7]

One of the Playfair Collection differential thermoscopes resembles, in some respects, Leslie's photometer, the other his pyroscope. Thomas Charles Hope demonstrated them to his chemistry class, and references to them can be found in the notes which he used for lecturing, for instance, in his lectures on 'Incandescence', he taught:

"Mr L has constructed an instrument to measure the intensity of these invisible [calori] fric emanations —
He calls its a *Pyroscope* —

It is the Differentl Thermr with one ball left clear & the other well coated with silver or Gold leaf —
Exh —".[8]

In his 'List of Specimens' (that apparatus to be brought out for his lecture demonstrations) again under the heading 'Incandescence', has been written: "Differential with *one ball Gilded*".[9]

However, in some respects the Playfair Collection instruments are rather different from those described by Leslie and those surving elsewhere which are thought to have been by the instrument making firms of Cary in London or Miller and Adie in Edinburgh under Leslie's guidance.[10] The major variations are that in Leslie's photometers and pyroscopes, the bulbs are brought close together (so that their local environment is as similar as possible) and the bulbs are of small diameter (Leslie mentions four to seven-tenths of an inch). Neither of the Playfair Collection instruments have scales. It is conceivable that Hope had these instruments specially constructed on a large scale and with the bulbs separated so that they would be visible to his very large classes. It was later mentioned that "To be visible to a class of 500 Students, his [Hope's] experiments required to be performed on a very large scale"[11] and in a note in his diary following a visit to his lectures in May 1818, William Allen remarked "The bulbs of his differential thermometer are nearly two inches in diameter, and his apparatus, in general, is upon a large scale."[12]

It is possible that the photometer and pyroscope were either early, experimental models produced by Leslie, or their constructor may have drawn inspiration from the 'co-rediscoverer' of the differential air thermometer,

Count Rumford (Benjamin Thompson), who in his paper read to the Royal Society in February 1804 said:

> "The instrument ... I shall take the liberty to call a *thermoscope* ... Like the hygrometer of Mr Leslie (as he has chosen to call his instrument), it is composed of two glass balls, attached to the two ends of a bent glass tube; but the balls, instead of being near together, are placed at a considerable distance from each other."[13]

Rumford also mentioned elsewhere that the glass bulbs should be 1½ to 1¾ inches in diameter.[14]

NOTES AND REFERENCES

1. W E Knowles Middleton *A History of the Thermometer* (Baltimore 1966) 25.
2. John Leslie *A Short Account of Experiments and Instruments depending on the Relations of Air to Heat and Moisture* (Edinburgh 1813) 48 ("I was led to the invention of the *Differential Thermometer,* an instrument extremely simple in its construction, and of singular utility in various delicate physical researches").
3. Letter of John Leslie to the editor of *The Caledonian Mercury,* dated 6 September 1812; reply by 'ME' written to the editor of the *Morning Chronicle,* dated 19 September. Both letters have been reprinted on a single sheet, see Edinburgh University Library MS Gen 269, envelope 43. See also 'Professor Leslie's Differential Thermometer invented by Professor Sturmius' *Edinburgh J Sci 2* 144 (1825).
4. John Leslie *An Experimental Inquiry into the Nature and Propogation of Heat* (Edinburgh 1804) 10.
5. *Ibid* 403.
6. *Ibid* 561.
7. Leslie *op cit* (2) 178.
8. Edinburgh University Library MS Gen 270 envelope 39.
9. Edinburgh University Library MS Gen 270 'List of Specimens'.
10. Leslie *op cit* (2) 178 ("The different Instruments and Machines described in this Tract, are to be had, of the most accurate and perfect construction, from Mr CARY, Optician, London, and from Messrs MILLER & ADIE, Edinburgh"). Instruments probably made by Miller and Adie (see D J Bryden *Scottish Scientific Instrument Makers 1600-1900* (Edinburgh 1972) 12, 13, 54) can be found in Teyler's Museum, Haarlem (see G L'E Turner and T H Levere *Martinus van Marum ... Volume IV. Van Marum's Scientific Instruments* (Leyden 1973) 263, item 210) and at the Royal Scottish Museum Edinburgh (acquired from the Department of Natural Philosophy, Edinburgh, nos. DU 31-38). These instruments are 'meters' rather than 'scopes' (ie they have scales) and they are more competently constructed.
11. Alexander Grant *The Story of the University of Edinburgh 2* (Edinburgh 1884) 398.
12. *Life of William Allen, with Selections from his Correspondence 1* (London 1846) 355.
13. Benjamin (Count) Rumford 'An Enquiry concerning the Nature of Heat' *Phil Trans Royal Soc 94* 77 (1804); see Sanborn C Brown (ed) *The Collected Works of Count Rumford 1* (Cambridge Mass 1968) 348.
14. Sanford C Brown (ed) *The Collected Works of Count Rumford 2* (Cambridge Mass 1969) 25. Paper first read on 19 March 1804 at the Institute de France and published 'Description of a New Instrument of Physics' *Mémoires de la Classe des Sciences ... de l'Institut de France 6* 71 (1806)

12. Voltaic pile
[Early form of the Voltaic Pile. 1858.275.11]

Overall height 431, base 152 x 152 mm.

A wooden (mahogany) base supports a glass rod (original?) and two glass capillary tubes (replacements?) which enclose the pile comprising 56 silver crowns, 45 copper discs (later additions?) and 57 felt pads (replacements?). A zinc contact protrudes from the base of the pile and at the top the glass tubes are held together with a wooden disc through which they pass.

Figure 24 Voltaic pile

Details of the first practical piece of apparatus to produce an electric current were transmitted to England in a famous letter, dated 20 March 1800, sent by the inventor, Alessandro Volta, to Sir Joseph Banks, President of the Royal Society. It was published in *Philosophical Transactions* later that year.[1] Volta's pile was the outcome of apparatus developed in a long series of experiments on the physiological effects of combinations of metals.[2] It consisted of circular discs of copper or silver alternating with discs of tin or zinc, each pair being separated by pieces of card or leather soaked in water or other liquid. Interest in the new phenomenon of 'galvanism' spread with amazing speed. In an article dated July 1800, William Nicholson suggested further practical details for the construction of a voltaic pile:

> "Take any number of plates of copper, or which is better of silver, and an equal number of tin, or which is much better, zinc, and a like number of pieces of card, or leather, or cloth*, or any porous substance capable of retaining moisture. Let these last be soaked in pure water, or which is better, salt and water, or alkaline lees. The silver or copper may be pieces of money.†
>
> *Woollen or linen cloth appear to be more durable, and more speedily soaked than card.
>
> †Most of our philosophers have used half crowns for the silver plates. The zinc may be bought at 3d per lb at the White Lion in Foster Lane, and may be cast in moulds of stone or chalk. A pound makes twenty thick pieces of the diameter of half a crown, or 1.3 inches diameter."[3]

The Playfair Collection pile incorporates 56 crown pieces, of which 46, though corroded, can still be identified.[4] Because the silver pieces alternate with copper discs, the pile would provide a very weak voltage. It is possible that the copper was incorrectly added in a later restoration of the apparatus, though there is an early account of the construction of a pile by the surgeon and lecturer J C Carpue in which the same erroneous construction is propounded.[5]

The idea of using silver coins to construct a pile was not originally Nicholson's. Indeed, the idea of a pile itself pre-dated Volta's publication by seven years and interestingly enough the suggestion was made by John Robison, professor of natural philosophy at Edinburgh from 1774-1805, and editor of Joseph Black's *Lectures on the Elements of Chemistry.* In a letter published in 1793, Robison stated:

> "I had a number of pieces of zinc made of the size of a shilling, and made them up into a rouleau, with as many shillings. I find this alternation, in some circumstances, increases considerably the irritation, and expect, on some such principle, to produce a still greater increase."[6]

Robison maintained his interest after the announcement of Volta's pile, for in October 1800 in a letter to James Watt he wrote:

> "You would oblige me if you would, in some idle hour, put down your notion of the *modus operandi* of the Galvanic Influence in the experts of Nicholson, Carlile, and Cruikshanks — I hear that davy, Dr Beddoeses Man has discovered that one end of Volta's Pile detaches Hydrogen alone, and the other as invariably detaches Oxygen . . . My pile, consisting of 72 pieces of each Metal, is well fitted up I think, for a variety of experiments but I have not been able as yet to make much use of it."[7]

As an aid to dating voltaic piles, it has been suggested that those which have three vertical supports are of later date than those which have four.[8] This postulation is probably based on the evidence of the diagram which accompanied Volta's paper of 1800.[9] However, reference to the text itself indicates that any such dating procedure is spurious.[10] The Playfair Collection can be compared with other early voltaic piles, but the provenance of most of these is far from being good enough to suggest a chronological sequence, especially as the instrument was copied and adapted so widely over a short space of time.[11] On the basis that the Playfair Collection pile incorporates silver coins and that John Robison must have stimulated interest in Edinburgh from an early date, it is likely that it was constructed in 1800 or shortly afterwards. Amongst Thomas Charles Hope's manuscript notes are to be found two printed advertisements for galvanic instruments, one issued by C H Wilkinson of Soho Square, London, and the other by T Brewin of the Birmingham Philosophical Society, but it is unlikely that the voltaic pile came from either of these sources.[12]

NOTES AND REFERENCES

1. 'On the Electricity excited by mere contact of conducting substances of different kinds. In a letter from Mr Alexander Volta F.R.S.' *Phil Trans Royal Soc 90* 403 (1800).

2. J R Partington *A History of Chemistry 4* (London 1964) 6.

3. William Nicholson 'Account of the New Electrical or Galvanic Apparatus of Sig. ALEX VOLTA' *Nicholson's Journal 4* 179 (1801).

4. The crowns are as follows: three King Charles II of 1662, thirteen King Charles II of 1663, one King William III of 1686, 26 King William III of 1692, three Queen Anne of 1708, ten unidentifiable.

5. J C Carpue *An Introduction to Electricity and Galvanism* (London 1803) 85 ("In the year 1800, Signor Volta discovered that if a number of plates of copper and silver, and pieces of cloth . . . are piled on another . . . a shock can be given").

6. Richard Fowler *Experiments and Observations relative to the Influence lately discovered by M Galvani and commonly called Animal Electricity* (Edinburgh 1793) 173, letter from John Robison to Fowler dated 28 May 1793.

7. E Robinson and D McKie *Partners in Science* (London 1970) 358, letter 228.

8. W James King 'The Development of Electrical Technology in the 19th Century' *Contributions from the Museum of History and Technology Bulletin 228* (Washington DC 1963) 237.

9. Volta *op cit* (1), plate opposite page 430.

10. *Ibid* 417 ("Dans la Fig 2de, mmmm, sont des montants ou baguettes, au nombre, de trois, quatre, ou plus, qui s'elevent du pied de la colonne, et renferment, comme dans un cage, les plateaux ou disques . . .").

11. Many relics of Volta, including a number of piles, were destroyed by the fire at the Tempio Voltiano, Como, Italy, in July 1899 (see Francesco Somiani *Tempio Voltiano* (Como 1928) and Bern Dibner *Alessandro Volta* (New York 1964) 104). The Museo Nationale della Scienza e della Tecnica, Milan, has reproductions of two piles associated with Volta, one held together by four thin rods of wood bound at the top (see Orazio Curti *Museoscienza* (Milan 1971) 165 and Henri Michel *Scientific Instruments in Art and History* (London 1967) 200, plate 103). The Museum of the Wellcome Institute of the History of Medicine, London, has a pile rescued from the Como conflagration. It has 32 discs of copper and zinc separated by felt pads, a wooden base and three glass supporting rods (inv no. 19/1950). The Royal Institution, London, possesses a pile presented by Volta to Michael Faraday. The discs are separated by three glass rods. A facsimile of this pile exists in the Science Museum, London (inv no. 1930-728). A pair of more sophisticated, very large, and perhaps later piles survive at Dickinson College, Carlisle, Pennsylvania, USA, of which a facsimile has been made for the National Museum of History and Technology, Smithsonian Institution, Washington (reg no. USNM 315049). The pile is supported by three glass and four wooden rods (see King *op cit* (8) 237, fig 4). A copper and zinc pile, separated by discs of parchment (possibly of 1821 or 1839) is in the collections at Transylvania University, Lexington, Kentucky (see Leland A Brown *Early Philosophical Apparatus at Transylvania College* (Lexington 1959) 7, 10, 86, fig 83). Another later example was acquired by Martinus van Marum and remains at Teyler's Museum, Haarlem, The Netherlands. It consists of a double pile of square plates held on the base by a central rod, clamp and screw (reg no. 575) (see G L'E Turner and T H Levere *Martinus van Marum Life and Work Volume IV. Van Marum's Scientific Instruments* (Leyden 1973) 343, fig 299). A late, decorative, apothecary's voltaic pile is to be found at the Pharmaziehistorisches Museum, Basel (see Pharmaziehistorisches Museum, Basel *Die Sammlung. Darstellung alter Arztinstrumente, Apotheker –Gefässe . . .3* (Basel 1974) 33, fig 17).

12. Edinburgh University Library MS Gen 269, envelope 54.

13. Sturgeon cell
[Cell of a Daniell's Battery. 1858.275.12]

Overall height 228, diameter of outer vessel 76mm.

The outer vessel consists of a copper cylinder with a base, painted black on the outside, and containing a copper cylinder concentrically fixed inside it. A zinc cylinder of intermediate diameter with a wire handle and three wooden insulating feet drops in between the copper cylinders. Two turned wooden cups attached by wires to the outer copper and zinc cylinders, when filled with mercury, make provision for electrical contacts.

Figure 25 Sturgeon cell

This cell was originally misidentified as a Daniell cell, which was first described in 1836 by John Frederick Daniell, professor of chemistry at King's College, London, from 1831 to 1845.[1] The Playfair Collection cell is in fact that developed in 1824 by William Sturgeon, an itinerant lecturer on science and a friend of Daniell.[2] In the 1824-25 session, Sturgeon was awarded the large silver medal and thirty guineas by the Society for the Encouragement of Arts, Manufactures and Commerce (now The Royal Society of Arts) for his Improved Electro-Magnetic Apparatus. Included in the apparatus was a cell, identical in operation to the Playfair Collection example, and described by the adjudicators as follows:

> "The battery is similar in construction to Professor Hare's calorimotor, and consists of two fixed, hollow, concentric cylinders of thin copper, having a moveable cylinder of zinc placed between them. Its superficial area is only 130 square inches and it weighs no more than 1 lb 5 oz. It is moveable on an upright metallic rod, like a laboratory lamp, and may therefore be adjusted to any convenient height... A further advantage of the construction is, that after every experiment the liquor may be returned to the jug, while preparations are making for the next, by which the battery, when wanted, is in a state of high activity, and is undergoing no deterioration in the interval between one experiment and another.
> A further point of novelty is, that Mr Sturgeon has very judiciously chosen to have a small galvanic power, assisted by a strong magnetic power, rather than the reverse, as is usually the case."[3]

A testimonial in Sturgeon's favour by Olinthus Gregory, professor of mathematics at the Royal Military Academy, Woolwich, makes it clear that Sturgeon developed the cell himself:

> "He [Sturgeon] wishes to solicit the patronage of your valuable Institution to a new battery of his invention, which, while it is very far smaller than any other of which I have heard, is very efficacious, and at the same time remarkably calculated to facilitate the experiments".[4]

The cell invented by Sturgeon[5] was advantageous in two respects. First, the anode was removable. Electrodes of simple cells suffered from deposition of gases on their surface formed by chemical action, and this prevented interaction between electrode and electrolyte; it was later termed 'gaseous polarization'. The anode could be reactivated in Sturgeon's cell by removing and cleaning it. Secondly, Humphry Davy had shown[6] that to increase the quantity of electricity (ie the current which produced the "strong magnetic power" referred to by Sturgeon) it was necessary to increase the area of the electrodes. An early improvement was suggested by W H Wollaston and acted on by J G Children: this was to increase the area of the cathode by folding a plate of copper in two, the zinc anode being inserted in between.[7] Robert Hare later suggested[8] that all the cathodes and all the anodes be joined together thus forming one pair of electrodes. Such a cell, which would produce a large current at low voltage, was called by Hare a 'calorimotor'; Sturgeon's cell was compared with it as it similarly had a large electrode surface for its overall dimensions.[3]

Sturgeon's cell does not seem to have been widely used, and most contemporary accounts of available electrical cells omit it. It was, however, described by D B Reid in 1839[9] and H M Noad in 1855.[10] The reason for its lack of wide popularity is probably that at the time cells were being developed at a rapid rate, and undoubtedly Daniell's cell later proved better for most purposes. The Playfair Collection cell shows slight differences from Sturgeon's original description and illustration[3]: the wooden cups to contain mercury are attached to different parts of the cell, and there is no provision to support the cell on a retort stand.

Sturgeon's cell was sold by the firm of Watkins and Hill of 5 Charing Cross, London who called it "Sturgeon's Cylindrical Battery", and described it as follows:

> "It consists of two fixed hollow concentric cylinders of copper having a moveable cylinder of zinc placed between them: one great advantage which this construction possesses is, that after every experiment the zinc cylinder may be taken out, and the oxide which is formed on its surface removed; the battery being thus restored to its original state of activity. 5s., 7s.6d., 10s.6d., to £1 5s."[11]

NOTES AND REFERENCES

1. J F Daniell 'On Voltaic Combinations' *Phil. Trans Royal Soc 126* 107 (1836); this letter from Daniell to Michael Faraday describes a galvanic cell consisting of a copper cylinder containing copper sulphate solution inside which was a piece of ox-gullet enclosing a rod of amalgamated zinc in dilute sulphuric acid. In a later letter, Daniell recommends the use of a thin earthenware tube in place of the ox-gullet (J F Daniell 'Further Observations on Voltaic Combinations' *Phil Trans Royal Soc 127* 141 (1837)). The Playfair Collection cell bears no functional connection with a Daniell cell. There is, it must be admitted, a superficial visual similarity between it and plate 9 in Daniell's 1836 paper and it is possible that the removable zinc anode was mistaken for a porous earthenware container of a Daniell cell.

2. J R Partington *A History of Chemistry 4* (London 1964) 685.

3. W Sturgeon 'Improved Electro-Magnetic Apparatus' *Trans Soc Arts 43* 37 (1824).

4. *Ibid* 43.

5. It should be noted that the cell discussed here is not the only cell invented by Sturgeon, and hence ambiguity in nomenclature must be avoided. 'Sturgeon's battery' is the name given to a cast iron cell developed by him: see *Ann Electr 5* 66 (1840) and William Sturgeon *Scientific Researches in Electricity* (London 1850) 183.

6. Humphry Davy 'Some Chemical Agencies of Electricity' *Phil Trans Royal Soc 97* 1 (1807).

7. J G Children 'An Account of some Experiments with a large Voltaic Battery' *Phil Trans Royal Soc 105* 363 (1815).

8. Robert Hare 'A New Theory of Galvanism' *American J Science 1* 413 (1819).

9. D B Reid *Elements of Chemistry* 3rd ed (Edinburgh 1839) 760. Reid was assistant to Thomas Charles Hope from 1828 to 1832 and may have had first hand experience of this particular cell.

10. H M Noad *Manual of Electricity 1* (London 1855) 251.

11. *Descriptive Catalogue of Voltaic and Thermo-Electric Instruments and Apparatus Constructed and Sold by Watkins and Hill* (London nd), copy in the Museum of the History of Science, Oxford. The firm of Watkins and Hill was in existence over the period 1806 to 1856 (after which it was absorbed by Elliot Brothers).

14. Voltameter
[Early form of Apparatus for decomposing water by the electrical current. 1858.275.13].

Overall height 228, diameter of cup 76mm. The volume measured by the five major divisions of the scale appears to be arbitrary: they measure 1.9 ± 0.3ml on one limb and 2.0 ± 0.2ml on the other (equivalent to 30 ± 4 and 31 ± 3 grains).

A glass tube branches into two limbs into each of which platinum wire electrodes are sealed; this is suspended above a turned wooden (yew) cup with a wooden support holding the tube in position. An ivory scale graduated into five major divisions, each of which is subdivided into ten parts, rests in between the limbs.

Figure 26 Voltameter

Voltameters were developed to demonstrate the decomposition of water and aqueous solutions by means of an electric current and to measure the volumes of gases evolved.

Before sources of continuous electrical current became available (ie before Volta's description of the pile was published in *Philosophical Transactions* in 1800[1]) a number of experimenters noticed that gas was evolved when static charges of electricity were discharged through water.[2] One of the first experiments using the voltaic pile was the investigation of effects caused by the passage of a current through water. Johann Wilhelm Ritter of Jena may have anticipated the well-known electrolysis experiments of William Nicholson and Anthony Carlisle by two years[3]: a letter to William Babington from a Dr G M of Freiburg, dated 17 December 1800, comments: "The principal galvanic discoverer here is a young man, called Ritter, at Jena: about two years since he published the result of his almost innumerable experiments in which he established all its laws and anticipated almost all the newer experiments."[4] The letter describes a U-tube fitted with corks through which gold wires, attached to a voltaic pile, protruded. Nicholson and Carlisle announced their results in a paper dated 2 May 1800[5] and similar experiments were reported by William Cruickshank[6] and William Henry[7] soon afterwards.

The first apparatus which could be used to measure the volume of an evolved gas may have been constructed by E G Robertson in Paris in 1800[8], though Wilhelm Ostwald, in his history of electrochemistry[9] considered that Robertson's apparatus had not been built and that the first water voltameter (though it was not called that until later, see below) was produced by P L Simon.[10]

The development of the design of voltameters from this early period, until the detailed descriptions of Michael Faraday in the 1830s, is not clear. Sometimes a U-shaped glass tube with corks through which the electrodes were passed was used, in which case no trough for the liquid being electrolysed was needed.[11] Another design had separate tubes covering each electrode, the tubes standing in a wineglass of liquid.[12] Few early diagrams show scales or graduated tubes, but most reports stated what proportion one gas bore to the other. Perhaps the earliest apparatus offered commercially was advertised by the London instrument maker and dealer Frederick Accum in 1805.[13]

Michael Faraday described and illustrated five forms of apparatus in 1834 in which he experimented on the electrolysis of acidulated water, two of which appear to have graduated collecting tubes.[14] He showed that the instrument could be used to measure the quantity of electricity, and called it a *volta-electrometer.* In his collected works *Experimental Researches in Electricity* published in 1839, the name has changed to its current form, *voltameter.*[15] However it seems that J F Daniell was responsible for coining this word; in a letter to William Whewell of 9 January 1836, Faraday mentions: "Perhaps you remember I gave the name of Volta-electrometer to the instrument I used in determining the definite action. . . Daniell advises calling it Voltameter which sounds shorter and well; is there any objection?"[16]

The Playfair Collection voltameter almost certainly predates the Faraday-type instruments (a number of which survive in the Royal Institution, London[17]). An early, but undated, voltameter with two separated, ungraduated tubes (marked 'H' and 'O') remains in Teyler's Museum, Haarlem.[18] Two early voltameters, both of which originally had separate collecting tubes (though only one remains) and which have electrodes protruding through the bases of glass troughs, survive in the Department of Chemistry at the Science Museum, London.[19] The Playfair voltameter would appear to be a rather early and unusual instrument.

NOTES AND REFERENCES

1. 'On the Electricity excited by mere contact of conducting substances of different kinds. In a letter from Mr Alexander Volta FRS' *Phil Trans Royal Soc* 90 403 (1800).

2. Giambatista Beccaria *A Treatise upon Artificial Electricity* (London 1776) 251, and see J R Partington *A History of Chemistry 4* (London 1962) 4; Beccaria (1682-1766), professor of medicine anatomy and chemistry at Bologna, was probably the first to notice the effect. In the late 1780s Martinus van Marum conducted a series of experiments on the calcinations of metals in water by means of electric sparks (see R J Forbes *Martinus van Marum Life and Work Volume I* (Haarlem 1969) 221). In 1789, Paets van Troostwijk and Johan Deiman published details of an experiment showing that hydrogen and oxygen were evolved when a Leiden jar is discharged through water (see *Obs Phys* 35 369 (1789)). These experiments were confirmed by George Pearson (1751-1828), physician at St George's Hospital, London, and a pupil of Joseph Black, in 1797 (see *Nicholson's Journal 1* 241 (1797)).

3. J W Ritter 'Versuche mit Volta's Galvanischer Batterie' *Voigt's Magazin für den neuesten Zustand der Naturkunde* 2 356 (1800).

4. W Babington 'On the State of Galvanism and other scientific Pursuits in Germany' *Nicholson's Journal 4* 511 (1801).

5. W Nicholson and A Carlisle 'Account of the new Electrical or Galvanic Apparatus of Sig Alex Volta' *Nicholson's Journal 4* 179 (1801). William Nicholson, an official of the East India Company, edited the journal which (popularly) bears his name. Sir Anthony Carlisle was a surgeon and professor of anatomy.

6. W Cruikshank 'Account of some important Experiments in Galvanic Electricity' *Nicholson's Journal 4* 187 (1801).

7. W Henry 'Experiments on the chemical Effects of Galvanic Electricity' *Nicholson's Journal 4* 223 (1801).

8. E G Robertson 'Sur la Fluide Galvanique' *Ann Chim* 37 132 (1800).

9. Wilhelm Ostwald *Electrochimie ihre Geschichte und Lehre* Leipzig 1896) 288.

10. P L Simon 'Beschreibung einer neuen galvanisch-chemischen Vorrichtung' *Ann Phys* 8 22 (1801).

11. Ostwald *op cit* (9) 160 (Ritter's apparatus).

12. John Cuthbertson *Practical Electricity and Galvanism* (London 1807) 256, experiment 204, figure 108; John Bostock *An Account of the History and Present State of Galvanism* (London 1818) figure 8. The Playfair Collection voltameter with its Y-shaped tube does not appear to have been described.

13. Frederick Accum *Catalogue of Chemical Preparations, Apparatus and Instruments* (London 1805) 19: "Galvanic Apparatus, for the rapid decomposition of water, 5s to 7s. . . Ditto, for obtaining the gases separate 12s 6d". Accum's apparatus was probably similar to that illustrated by J C Carpue, a London surgeon and lecturer, who in his *An Introduction to Electricity and Galvanism* (London 1803) (referred to by Accum in the above *Catalogue. . .*) describes two kinds of apparatus, one in which the gases are collected together (page 105, plate III figure 5) and the other in which the gases are collected in two tubes which stand on a plate in a beaker of water the electrodes passing through the sides of the beaker and protruding up into the tubes (pages 97, 105, plate III figure 8).

14. M Faraday 'On a new Measurer of Volta-electricity' *Phil Trans Royal Soc* 124 85 (1834).

15. Michael Faraday *Experimental Researches in Electricity 1* (London 1839) 206.

16. L Pearce Williams (ed) *The Selected Correspondence of Michael Faraday Volume I 1812-1848* (Cambridge 1971) 301 (letter 196). William Brande first published the word 'voltameter'. — see W T Brande *Manual of Chemistry* 4th ed (1836) 263: "Mr Faraday has constructed several instruments for this purpose which he terms Volta-electrometers or Voltameters."

17. They bear the following register numbers: 165 (or D2), 169 (or D7), 170 (or D8) and 171 (or D5). The Playfair voltameter probably also predates an example (no longer extant) acquired by the Department of Natural Philosophy, Edinburgh, circa 1833, which had been supplied by the firm of Watkins and Hill, London (see Royal Scottish Museum MS 'Natural Philosophy Museum — Register of Apparatus' item E70).

18. G L'E Turner and T H Levere *Martinus van Marum. . . Volume IV. Van Marum's Scientific Instruments* (Leyden 1973) 345, item 340, figure 301.

19. Inv Nos 1937-156 and 1946-48.

15. Instantaneous light box
[Oxymuriate match box. The matches were tipped with chlorate of potash & sugar, which produced a light when dipped in the bottle containing sulphuric acid. The apparatus was in use before lucifer matches. 1858.275.15]

Box 63 x 38 x 76 (high), diameter of bottle 32mm.

The tinned iron box painted red internally with an external *moiré metallique* finish contains two compartments, one of which is itself hinged. A small glass bottle with ground glass stopper and cap (for containing the sulphuric acid) fits into the other compartment.

16. Instantaneous light box
[Another of a different form. 1858.275.16]

Box 98 x 57 x 43 (high), bottle 26 x 26 x 35mm (high)

Tinned iron box painted red internally and black externally, with a yellow flame-like frieze on the lid (which is hinged). The box contains three compartments to house a small square cross-section glass bottle with a ground glass stopper, a bundle of wooden splints and a candle fitted with an iron ring holder for fixing inside the box in an upright position.

Figure 27 Instantaneous light boxes: object 15 (right), object 16 (left)

The instantaneous light box or oxymuriate match box was introduced to Great Britain from France soon after 1810.[1] It consists of a small box, usually of tinned iron, containing a small stoppered bottle containing sulphuric acid and a number of wooden splints, with a match head consisting of a mixture of potassium chlorate and sugar made into a paste, gum being added to harden the mixture. When a match is dipped into the acid and withdrawn, it bursts into flame. The instantaneous light box was one of a number of devices for fire-making which were current until about 1835 when all had been ousted by the friction match in its various forms. This was first produced in Britain by

99

John Walker of Stockton-on-Tees in 1827 though not widely used until lucifers and congreves became available from 1829 and 1832 respectively.

It has been suggested that boxes which contain inner compartments with lids are of relatively late construction.[2] This provision was made to house the matches and was intended to prevent a stray drop of acid falling on them accidentally and igniting the whole lot at once. Both Playfair Collection examples have such inner lids. It can be assumed that both instantaneous light boxes were acquired by Thomas Charles Hope between about 1810 and 1830. The notes from which he lectured (under the heading 'Chloras Potassae 2d') contain references to the apparatus.[3]

NOTES AND REFERENCES

1. Miller Christy *The Bryant and May Museum of Fire-Making Appliances* (London 1926) 12, 94. Included in the Bryant and May Collection (now at the Science Museum, London) is a photograph of object 15 in the Playfair Collection (see item 1268, *ibid* 96).
2. *Ibid* 94.
3. Edinburgh University Library MS Gen 269, envelope 63 ("It is now common to have an apparatus for getting light... They are in the Shops for Sale"). See also MS Gen 270 ("List of Specimens") which includes "Matches with Compn Chlor. Pota".

17. Argand lamp
[Argand Oil Lamp. 1858.275.17]

Overall length 260, diameter of oil reservoir 81, diameter of wick holders 38 (1½ inches), 17 mm.

The lamp (the stand for which is missing) comprises a reservoir of cylindrical form with a sloping base connected to the lamp at its lowest edge by two converging tubes to feed the two wicks (housed in hollow concentric tubes) with oil (the wick raising mechanism is missing). The reservoir contains a concentric cylindrical vessel terminating in the domed cover (which acts to regulate the oil flow) fitting closely inside the outer cylinder. The lamp is japanned externally and is decorated with gold painted motifs.

Figure 28   Argand lamp

Oil lamps (or 'lamp furnaces') and spirit burners were, along with portable or built-in coal, coke or charcoal furnaces, the main sources of heating for laboratory experiments until the introduction of efficient gas burners in the middle of the 19th century. One of the first efficient oil lamps was that developed from 1780 by Francois-Pierre Ami Argand to provide better illumination than hitherto. By incorporating a tubular wick, a flow of air was drawn up through the centre which resulted in more complete combustion and a brighter flame. The English patent was taken out in 1784.[1] The early history of the lamp has been extensively traced.[2]

Initially the Argand lamp was not generally used for heating purposes in laboratories. However, referring to the lamp in a letter to James Watt dated 11 August 1785, John Robison wrote "I can truly say that I made a Lamp on a principle nearly the same, in the year 1767, but with a common wick."[3] He reinforced this statement in a note to his edition of Black's Lectures:

> "The sure way of producing a more complete inflammation is to prevent a great quantity of surface from flaming at once. Thus, a very small wick is found to burn without smoke or soot; and Argand's lamp, which admits a current of air into the centre of the flame, produces the same effect in an eminent degree. I employed such a lamp in 1767-8. It consisted of two concentric circles of wicks of rush pith stuck on pins. The air came up in the centre, and between the circles. This was the lamp of a distilling apparatus, and stood on the laboratory table of Glasgow college for two years. Quitting all chemical pursuits in 1769, I thought no more of it."[4]

The Argand lamp with two wicks, for heating rather than illumination, was reintroduced by Frederick Accum in 1804. Describing it, he said:

> "The chief improvement of this lamp consists in its power of affording an intense heat by the addition of a second cylinder added to that of the common lamp or Argand. This additional cylinder incloses a wick of one inch and a half in diameter, and it is by this ingenious contrivance which was first suggested to me by Mr Webster, that a double flame is caused."[5]

In his catalogue of 1817, Accum claimed that in his version "the power of producing heat is augmented to more than three times that afforded by the ordinary lamp of Argand."[6] His "Lamp Furnace with concentric wicks" cost from £4 14s 6d to £5 15s 6d.

It is not clear whether Joseph Black demonstrated the Argand lamp to his classes. He certainly was informed about the introduction of the lamp in 1784.[7] Argand's patent refers only to the tubular wick and air supply, and a number of modifications were proposed soon after its introduction. One of these was by Peter Keir of Edinburgh, described in his patent specification as 'mathematical instrument maker'.[8] He was probably known to Black: John Robison spoke of him in a letter to James Watt.[9] However the painted decoration on the lamp is most likely to date it to the first quarter of the 19th century. It shows close similarities to the gold decoration on an Argand lamp at the Berzelius Museum at the Royal Swedish Academy of Science which was probably acquired by Jöns Jacob Berzelius in Paris in the years 1818-1819.[10] A similar lamp to the Playfair example with separate reservoir and burner was described and illustrated in 1824.[11]

NOTES AND REFERENCES

1. Ami Argand 'Lamp, producing neither smoke nor smell, and giving more light than any before known' English patent 1425, 15 March 1784.

2. Michael Schrøder *The Argand Burner* (Odense 1969).

3. E Robinson and D McKie *Partners in Science* (London 1970) 148, letter 103 (Robison to Watt, 11 August 1785).

4. Joseph Black *Lectures on the Elements of Chemistry... Published by John Robison* (Edinburgh 1803) 539. See also ref. (9).

5. Frederick Accum 'Description of a chemical lamp, with double Concentric Wicks' *Nicholson's Journal 8* 266 (1804).

6. Frederick Accum *Descriptive Catalogue of the Apparatus & Instruments Employed in Experimental and Operative Chemistry* (London 1817).

7. Edinburgh University Library MS Gen 873 volume 2, ff 195-6 letter from John Grieve to Joseph Black, dated Hertford, 8 October 1784 ("As Mr Argaund (sic) the inventor of the New Lamps is come to London, and is joined with Mr Parker in a Patent for them you either do already or soon will know them").

8. Peter Keir 'Raising the oil supply in lamps' English Patent 1585, 29 January 1787.

9. Robinson and McKie *op cit* (3). Robison, writing to Watt, asked whether he could persuade Matthew Boulton to manufacture Peter Keir's new form of the lamp. Keir was praised by Robison as being "a pretty good workman at the Lamp, and in Brass and hardwood turning he is exceedingly neat. He has lately made an Electrical Machine for me in the most accurate and complete Manner and has a good notion of philosophical (not mathematical) instruments in general." The letter is of interest in that it is one of the very rare descriptions of a scientific instrument maker in Edinburgh in the 18th century and it refers to the "narrow demand" for his products in that city.

10. Sven Klemming 'Om Stockholms belysning, offentlig och privat, under perioden 1800-1850' *Daedalus, Teknisk Museets Arsbok 98* (1971), and private communication.

11. *An Explanatory Dictionary of the Apparatus and Instruments of... Chemistry* (London 1824) 136, plate 7 fig 6. See also Samuel Gray *The Operative Chemist* (London 1828) 94, plate 19 fig 56 ("An Argand's lamp... represented at Fig 56, is very frequently employed at present for evaporations, and similar operations which do not require any great heat.")

18. Lamp furnace
[Lamp Furnace. 1858.275.19]

Overall height 413, diameter of cylindrical body 120 mm.

The furnace is constructed from tinned iron; the burner, surrounded by an annular reservoir, is a single wick Argand lamp with a device by which the height of the lamp can be controlled. The lamp is surrounded by the body of the furnace which contains a hinged door and is supported on three legs. A flared upper portion is fixed to the top to accommodate the vessel to be heated which can be supported by six adjustable wire rods.

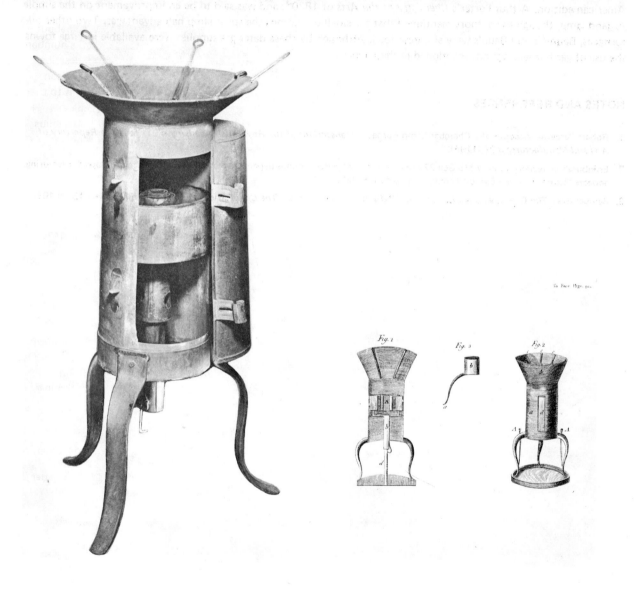

Figure 29 Lamp furnace

Figure 30 Lamp furnace, from Robert Perceval 'Account of a Chamber Lamp Furnace' *Transactions of the Royal Irish Academy* 4 [1791]

This oil-burning lamp furnace is a development of the Argand lamp (see Playfair Collection object 17) for heating purposes in laboratories. It was devised by Robert Perceval an Irishman who had graduated MD at the University of Edinburgh in 1780. In 1785 he was elected the first professor of chemistry at Trinity College, Dublin. On 5 March 1791 he read a paper 'Account of a Chamber Lamp Furnace' to the Royal Irish Academy (which he took an active part in founding in 1785).[1] This apparatus was said to overcome the problem of heat loss to which the Argand lamp was subject by surrounding the lamp with a cylindrical furnace body. This was surmounted by a cone-shaped section in which vessels were rested (called by Perceval the 'laboratory'). The lamp could be raised or lowered on a stem, fixed at different heights by means of a spring catch. Experiments conducted by Perceval on the furnace showed that too liberal a supply of air through the centre of the burner could result in a cooler flame.

The Playfair Collection lamp furnace is substantially that described by Perceval in 1791, though a tray, funnel and three iron spikes for supporting vessels are missing, and the dimensions are somewhat larger than those mentioned by him. The furnace was shown by Thomas Charles Hope to his chemistry class.[2]

Perceval's lamp furnace was discussed in some detail in Samuel Gray's *Operative Chemist* of 1828 and its equivalent American edition, Arthur Porter's *Chemistry of the Arts* of 1830[3], and was said to be an improvement on the simple Argand lamp, though the authors mentioned that for small operations the spirit lamp had advantages. Two other oil furnaces, Baumé's and Baup's were also described. Although by these dates gas supplies were available in large towns, the use of gas burners was not mentioned in these works.

NOTES AND REFERENCES

1. Robert Perceval 'Account of a Chamber Lamp Furnace' *Transactions of the Royal Irish Academy* 4 91 (nd); *The Repertory of Arts and Manufactures* 3 24 (1795).
2. Edinburgh University Library MS Gen 272 (within a pile of miscellaneous papers). Amongst the notes used by Hope for lecturing appears "Lamp furnace - Exhibit-Percivals Lamp furn Exht".
3. Samuel Gray *The Operative Chemist* (London 1828) 79; Arthur L Porter *The Chemistry of the Arts 1* (Philadelphia 1830) 101.

19. Retort stand, boss and three clamps

[Old form of Laboratory Retort Stand.] Original registration number 1858.275.21. The apparatus was disposed of by a Board of Survey of 4 April 1946, though not destroyed. It was readmitted to the Museum Register in 1974 and now has the number 1974.287.1-3.

Height of stand 477, base 178 x 140. Length of clamps 170, 170, 115mm.

The stand consists of a wooden (birch) rectangular base, in the corner of which is screwed a wooden rod. The boss, secured to the rod by a wooden screw, incorporates a short wooden rod which can be accommodated in holes drilled in the clamps. Two of the clamps have sprung wooden jaws adjusted by wooden screws. The third has a clamp which moves towards a fixed jaw, guided by two metal wires. The boss and the three clamps are of mahogany.

Figure 31   Retort stand, boss, two Berzelius holders, Sefstroem holder

Supports for chemical apparatus developed in a complicated manner over a long period. Supports and stands for use in distillation operations are illustrated in the earliest graphic representations of chemical experiments.[1] Descriptions of such stands are, however, rare. Occasionally they are mentioned in captions to figures accompanying articles describing chemical experiments.

The stands offered by the London dealer Frederick Accum in his 1817 Catalogue were of brass or iron, with sliding rings and with or without adjusting screws.[2] There are no illustrations of these in the Catalogue, but figures accompanying other works by Accum[3] show the stands to have substantial tripod bases. On the other hand, John Joseph Griffin (of the Glasgow firm Richard Griffin and Company) offered, in 1838, retort stands with flat rectangular bases with brass, iron or wooden (stained black) rods.[4] A variety of clamps is illustrated and described by J J Griffin, named after those whom he considered to have introduced them. These include retort holders of J L Gay-Lussac, N G Sefstroem and J J Berzelius, and a cylinder holder of J G Gahn. Such clamps, in a modified form, were sold by Griffin, together with a wooden retort stand, and the ' support for glass tubes etc' (see Playfair Collection object 22). The whole set, which included one of each item (except the Gay-Lussac holder) was sold for 14 shillings, but if varnished with copal cost an extra 1s 6d.[5]

It is quite likely that J J Griffin was responsible for the firm of Richard Griffin and Company offering these new kinds of stands for sale in Britain. C Greville Williams wrote in 1857:

"Mr Griffin, to whom the public is indebted for the introduction from the Continent of an immense number of instruments, until then unknown in England, has described a vertical clamp well adapted for the support of small tube retorts, and several other pieces of apparatus."[6]

In 1828 and 1829 Griffin had studied abroad first in Paris and then under Leopold Gmelin at Heidelberg.[7]

The Playfair Collection retort stand and holders are nearly identical to those described in Griffin's 1838 catalogue and were almost certainly supplied by the firm of Richard Griffin and Company. The height of the rod (18 inches) is the same as that advertised. In addition the Playfair Collection includes the boss (which Griffin called the Nutt), the Sefstroem holder (which closely resembles the "modification of this apparatus, as it is now made for sale in Glasgow"[8]) and two Berzelius holders.

NOTES AND REFERENCES

1. Very many types of support for chemical apparatus are shown in the figures in R J Forbes *A Short History of the Art of Distillation* (Leiden 1970).

2. Frederick Accum *Descriptive Catalogue of the Apparatus and Instruments Employed in Experimental and Operative Chemistry* (London 1817) 34. This catalogue is printed at the end of Frederick Accum *Chemical Amusement* (London 1817), and Frederick Accum *A Practical Essay on Chemical Re-Agents or Tests* (London, 1816).

3. Frederick Accum *Chemical Amusement* 3rd ed (London 1818) and Frederick Accum *A Practical Essay on Chemical Re-Agents or Tests* 2nd ed (London 1818).

4. J J Griffin *Chemical Recreations* 8th ed (Glasgow 1838) 296, 297. The Glasgow firm of scientific instrument makers, Richard Griffin and Company, first appears in the Glasgow Post Office Directory of 1820. Richard Griffin's brother John Joseph Griffin went into partnership with the company in 1832 shortly before Richard's death. J J Griffin moved to London in 1848, setting up there as a chemical instrument supplier and manufacturer. The firm of Richard Griffin and Company was the only substantial firm of chemical apparatus makers in Scotland in the first half of the 19th century.

5. *Ibid* 297.

6. C Greville Williams *A Handbook of Chemical Manipulation* (London 1857) 159.

7. Typescript biographical notes supplied by Charles Griffin and Company Limited, Publishers, London WC2.

8. Griffin *op cit* (4) 282.

20. Heating bath (?)

[Square Retort Stand enclosing a heating bath. 1858.275.22]

Height of bath 368, height of pillar 356, base 438 x 146, dimensions of bath 102 x 102 x 190 mm. (deep).

A square wooden (fir) casing (open to one side at the base) houses a (corroded) iron bath in which a notch has been removed from the upper side which faces a wooden (mahogany) pillar screwed into the opposite side of a rectangular wooden base.

Figure 32   Heating bath

The function of this crudely constructed apparatus is not certain. It seems likely that it was used for gentle heating, a lamp being inserted beneath the metal bath. The wooden sides supporting the bath are scorched. It may have resembled a stove constructed by Jean D'Arcet, described as "a four-sided chest, made of very dry wood . . . the temperature could not be raised above the heat of boiling water, even after several hours."[1] The bath may have contained sand, or more likely, water, for use as a *bain-marie*.

The notch removed from the side of the bath may have allowed the neck of a small retort to protrude towards the upright column. This latter may have originally supported a clamp to hold the neck of a flask or tube being heated.

REFERENCE

1.   A L Porter *Chemistry of the Arts 1* (Philadelphia 1830) 105.

21. Adjustable table support (?)
[Support for glass tubes &c. 1858.275.23]

Overall height 330, base 210 x 152 mm.

Two pairs of square cross-section wooden (mahogany) rods held together at their bases by iron clamps and wing nuts are mounted on a wooden base (mahogany), one of each pair of rods being secured by a hinge. The rods are lined on their inner surfaces with green felt.

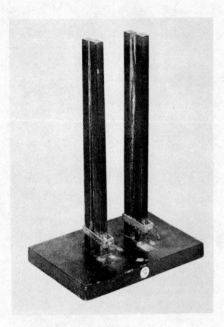

Figure 33 Adjustable table support

This support was probably used in conjunction with a table (now missing) whose supports would run in between the vertical members, clamped by the wing nuts. The height of the table could thus be vertically adjusted.

An early variety of adjustable vertical support (but sliding in grooves) was illustrated by Peter Woulfe in his 1768 paper 'Experiments on Distillation'.[1] However the Playfair Collection example closely resembles an illustration of a clamp which appears on a chemistry class ticket issued by William Irvine in 1785 at Glasgow University.[2] In this illustration, the table is shown supporting a receiver in a distillation process (which, incidentally, is utilising a Black-type portable furnace).

A somewhat similar vertical clamp was illustrated in Greville Williams' *Chemical Manipulation*.[3] Later the firm of John Joseph Griffin and Son offered a modification described as "Vertical Clamp, for supporting tubes, small retorts, &c . . . mahogany, French polished."[4] However, the example in the Playfair Collection has a rather early variety of wing nut and may date from the 18th century.

NOTES AND REFERENCES

1. P Woulfe 'Experiments on Distillation' *Phil Trans Royal Soc 57* 517 (1768).
2. Andrew Kent (ed) *An Eighteenth Century Lectureship in Chemistry* (Glasgow 1950), plate XIII.
3. C Greville Williams *Handbook of Chemical Manipulation* (London 1857) 160, fig. 136.
4. J J Griffin *Chemical Handicraft* (London 1866) 29.

22. Multi-purpose apparatus support
[Support for glass tubes &c. 1858,275.23]

Height 127, diameter 121 mm.

A cylindrical wooden (mahogany) stand supported on three wooden legs contains a circular hole, a circular conical hole, two arcs of circles cut into the circumference and a hole drilled a short distance into the side.

Figure 34   Multi-purpose apparatus support

This apparatus was designed and sold by Richard Griffin and Company and described by J.J. Griffin in his *Chemical Recreations* of 1838. It was a multi-purpose instrument holder incorporating the features from a variety of other supports for chemical apparatus. It was described as:—

> "THE TABLE. — This branch is rather of complex structure. It is a combination of the set of Berzelius's supports . . . The flat part is the table . . . This table is five inches in diameter, and serves to support the lamp furnace . . . and the gas light table . . . The crook . . . is supplied by two notches in the sides of the table, one adapted for tubes of an inch in diameter, the other for tubes of two inches in diameter, such as the condenser . . . Brass headed nails are inserted on each side of these notches, to afford the means of fastening the tubes to the table by means of string or thin metallic wire, slips of Indian rubber being put between the table and the tube to facilitate their adhesion. The pegs on the reverse of the table . . . serve to hold round bottomed vessels — basins or flasks. The table is pierced with two holes. One of these is cylindrical, and intended for the support of tubes placed horizontally when the table is fixed vertically . . . The other perforation is conical and is intended to support large funnels employed in the filtration of heavy masses of liquid."[1]

The description above is found in an appendix entitled "Additions to the Preceding Articles", is not found in the 7th edition of 1834, and so was introduced by Griffin between then and 1838.

REFERENCE

1. J J Griffin *Chemical Recreations* 8th ed (Glasgow 1838) 283. For notes on John Joseph Griffin, see Playfair Collection, object 19.

23. Still boiler
[Copper Still for careful distillations. 1858.275.24]

Overall height 642, diameter of lower cylinder 190, diameter of upper cylinder 223mm.

The boiler is constructed from two concentric copper cylinders (both of which are constructed with joins brazed with turret seams). A copper funnel and tube is attached to the domed surface of the upper cylinder; it extends halfway down into the boiler.

Figure 35 Still boiler

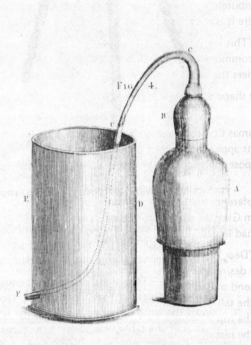

Figure 36 Black's still, from *The New Dispensary...* by Gentlemen of the Faculty at Edinburgh (Edinburgh 1786) plate 3

Distillation is a process whereby a more volatile component of a mixture of liquids is separated from the less volatile components by boiling, condensing the vapour on a cold surface and running the liquid into a receiver. It was one of the earliest chemical operations to be evolved. The still (in the western hemisphere) developed into two basic forms, the alembic (see Playfair Collection, object 56) and the retort (see Playfair Collection, objects 52-54). These were generally constructed from glass, earthenware, copper or pewter.[1] Those made from metal were separable into several parts, the boiler, the still head and the condenser. This still in the Playfair Collection consists of the boiler only. The tube and funnel joined to it were for replenishing the liquid being distilled so that the process would not have to be interrupted. Stills of the same basic design were available from the mediaeval period to well into the 20th century.[2] The Playfair Collection still could date from the occupancy of the Edinburgh chair by Black, Hope or Gregory and it could possibly be even earlier.

Black developed a slight modification to the condenser of the still, which was frequently of spiral form, passing through a barrel of cold water. Descriptions of Black's still frequently appear in notes taken by students attending his lectures. A particularly full account reads:

> "In the distillatio per ascensum, the vapour is allowed to follow its natural tendency, to arise to some height upwards and then is conveyed into a cold cavity to be condensed. For this, a variety of vessels have been contrived. The most ordinary is a Copper Still. That which I make use of varies a little from the common form. It consists of the body, containing the materials; and the head into wch. the vapour immediately rises. From this issues a pipe, which is directed to a side, and descends downwards again, in order to be inserted into the extremity of another pipe which descends into a large vessel of copper, containing a quantity of cold water. In the ordinary stills, the pipe is made serpentine, but that renders it difficult to be cleaned. And it is not necessary where the apparatus is small, and the vapours are very quickly condensed, provided there is a sufficient quantity of cold water around them. So I have the pipe made of a more simple form: it descends immediately into the vessel, then passes obliquely across the bottom to the opposite side, where it issues out, and where the vessel receiving the condensed matter is supplied to it."[3]

John Robison's published edition of Black's lectures does not contain such a complete account.[4]

A description and illustration of Black's form of still (fig 36) was published 'By Gentlemen of the Faculty at Edinburgh' in a 1786 edition of William Lewis's *New Dispensatory,* edited by Charles Webster and Ralph Irving, where it is mentioned:

> "This instrument is on the construction used and recommended by Dr Black, and varies a little from the common form. The Doctor finds it unnecessary that the pipe D should be made serpentine, which renders the cleaning of it very difficult and uncertain."[5]

The shape of the boiler is very similar to the Playfair Collection example.

Thomas Charles Hope discussed the use of stills in his lectures, and instruction to his assistant regarding their employment appear in his 'List of Specimens' to be prepared for his lectures: "To distil a larger quantity of water for other purposes, with the Small Copper Still, set it to work immediately after lecture."[6]

A reference to a specific still used by Black is made in a letter sent by him to James Watt in 1768 after he had moved from Glasgow to the Edinburgh chair of chemistry. Requesting that he be sent a number of pieces of apparatus which he had left behind in Glasgow, he wrote:

> "Dear Watt
> I desired Mr Hill to send you some things from the Laboratory which I must beg of you to pack up and send me by the Carriers — among these are an absurd sort of still and a tall head to it both of copper — the tall head you may sell as old Copper — the Still or Body, I wish to have opened above by taking off the top of it which is soldered on only with soft solder and that top is also to be sold as old Copper — the rest of the Body will serve me as a boiler and may be sent packed full of the other things."[7]

It is tempting to imagine that the still boiler in the Playfair Collection might be the very one referred to in this letter.

NOTES AND REFERENCES

1. The most comprehensive account of the history of distillation is given by R J Forbes *A Short History of the Art of Distillation* (Leiden 1948); see also F Sherwood Taylor 'The Evolution of the Still' *Annals of Science 5* 185 (1945).
2. *Ibid* 76, 85; Baird and Tatlock (London) Ltd. *Price List of Chemical and Bacteriological Apparatus* (London 1914) 438.
3. Chemical Society, London, MS E151i ('Lectures on Chemistry by Joseph Black' 6 volumes (1774?)) lecture 26.
4. Joseph Black *Lectures on the Elements of Chemistry... Published by John Robison* (Edinburgh 1803) 166, 299.
5. *The New Dispensatory... By Gentlemen of the Faculty at Edinburgh* (Edinburgh 1786) 57, plate 3. For the bibliographic complexities of this series of pharmacopoeias, see David L Cowen 'The Edinburgh Dispensatories' *The Papers of the Bibliographical Society of America 45* 85 (1951).
6. Edinburgh University Library MS Gen 270 'List of Specimens'.
7. E Robinson and D McKie *Partners in Science* (London 1970) 15, Letter 7 (Black to Watt, 31 October 1768).

24. Water-cooled condenser
[Liebig's Condensers. 1858.275.25]

Height 285, length of trough 450 mm.

A japanned tinned iron trough through which an iron tube runs, emerging at either end, is supported by two pairs of iron legs of unequal height, allowing the trough and tube to slope. An iron overflow tube is soldered to a point close to the upper edge of the trough, around which is painted a frieze of floral motifs in yellow.

25. Counter-current condenser
[Liebig's Condensers. 1858.275.25]

Length of inner tube 863, diameter 16, diameter of outer tube 63 mm.

Two pale-green glass tubes are held concentrically by brass mountings from which extend short brass tubes (for attachment to a water supply, allowing water to flow through the outer glass tube). The outer tube is fixed by a brass collar to an adjustable joint, attached to a brass pillar held in an iron base.

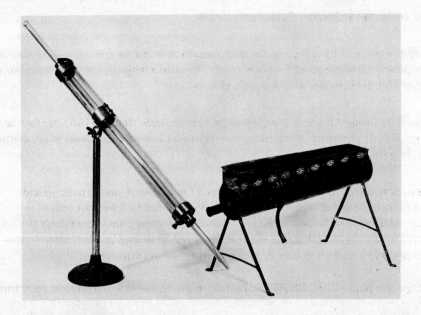

Figure 37 Water-cooled condenser (right), counter-current condenser (left)

In order to carry out efficient distillation it is necessary to have an effective method of condensing the vapour of the distillate. In the simple alembic and retort-type stills, the vapour is condensed on an air-cooled concave surface and run off. For liquids with high boiling points this was adequate, but for more volatile substances such as alcohol a further system of condensation was found to be desirable. Apparatus was used in the Middle Ages onwards which consisted of a spiral tube of earthenware or metal which was immersed in a reservoir of water (worm condenser). The water warmed up as the process proceeded and it was sometimes necessary to replace the hot water by fresh charges of cold water. This is the background to the development of the counter-current condenser (sometimes called the Liebig condenser, see below) in which a flow of cold water is maintained in an outer concentric tube in the opposite direction to the flow of the distillate in the inner tube.

The counter-current condenser was developed in the decade 1770-1780 by three independent workers or groups in Germany, France and Sweden. After the initial evolution of the apparatus interest waned and did not revive until about 1830, from which time it has been in constant use in the chemical laboratory. Of the two examples of water-cooled condenser in the Playfair Collection, one (object 24) is a proto-counter-current condenser and the other (object 25), is an early example of a true counter-current condenser.

The apparatus was first described by Christian Ehrenfried Weigel in his MD thesis published at Göttingen in 1771 which included a diagram of his counter-current condenser.[1] It consisted of two concentric tinned iron tubes. The condenser was fixed at an angle and water was poured through a funnel attached to the lower end of the outer tube; the distillate condensed in the inner tube and trickled down into a receiver. In the second volume of the thesis published two years later Weigel solved problems of corrosion and leakage by using a glass inner tube joined to the outer tube with a lute.[2]

Weigel's distillation apparatus was intended for laboratory use; the other two counter-current condensers developed at about the same time were intended for larger scale operations. A pamphlet, published in Paris in 1781, described a still to be used for the desalination of sea water. It was almost certainly written by the scientific communicator J Hyacinth de Magellan.[3] This stated that a counter-current condenser was first proposed in 1770; models were made in January 1773 after which, on the orders of the Minister of the Navy and Colonies, P-E Bourgeois de Boynes, it was constructed full size. The apparatus was then shown to several members of the Académie Royale des Sciences (including A-L Lavoisier). The condenser consisted of two square cross-section copper or white tin tubes with thin walls. The inner tube was suspended in the outer tube with small metal plates which did not inhibit the flow of water. The square section was preferred because of relative ease and cheapness of manufacture, and because of the larger surface area available for cooling.

The third independent inventor of the counter-current condenser was the Swede Johann Gadolin who published a description of the apparatus in 1791.[4] The outer tube was constructed of pine wood and the inner tube was oval in cross-section and made of sheet copper. Gadolin realised that he had been anticipated soon after his original publication[5] though his claim to independent discovery is credible because his father, Jacob Gadolin, described a similar, though not counter-current, condenser thirteen years earlier.[6]

Although it is not altogether clear who first produced a counter-current (or Liebig) condenser[7] one thing is certain: it was not Justus von Liebig who was not born until 1803. Liebig first illustrated the condenser in his 1832 paper on the preparation and properties of chloral.[8] By 1842 in the *Handwörterbuch der Chemie* (of which Liebig was co-author) a reference to the condenser appears which could be taken to imply that Liebig had first described it.[9] Thomas Graham referred to "the condensing tube of Professor Liebig" at about the same time.[10]

In order to date the two water-cooled condensers in the Playfair Collection it is necessary to discover from what time dealers were offering such apparatus: object 25 is certainly a competently made instrument, probably constructed by a professional maker. It seems that revival of interest in the apparatus coincided with increasing attention being paid to preparative organic chemistry (which required efficient distillation procedures). It is somewhat surprising, however, that Michael Faraday omitted the condenser in his third edition of *Chemical Manipulation* of 1842[11], though it had been described in 1839 by Thomas Charles Hope's erstwhile assistant, David Boswell Reid.[12] The counter-current condenser was also described by J J Griffin in 1838, though it was not offered by his firm of Richard Griffin and Company at that time; indeed, he suggests that "it is one which a chemist can very easily get made in any village where there is a blacksmith".[13] The firm did advertise an all glass apparatus (for the first time?) in 1866 when it is described as "Condenser tube, hard German, fitted in a glass envelope".[14] An imported apparatus had been offered eighteen years earlier by the instrument dealer Charles Button.[15]

Early surviving counter-current condensers can be divided into two types, those in which the cooling-water inlet is directly above the outlet, an arrangement made possible by leading the inlet through a thin tube alongside the outer tube to the lower end of the condenser[16], and those where the inlet and outlet are at opposite ends of the outer tube.[17] Those of the former variety are likely to have the outer tube constructed of metal. It is not clear whether one style preceded the other. The glass condenser in the Playfair Collection is of the latter type and it may date from 1830 to 1858. It is difficult to understand why Lyon Playfair discarded such a well-made, serviceable instrument.

The metal water-cooled condenser in the Playfair Collection is equally difficult to date. It does not appear to be of a type offered by instrument dealers. The *paterae* decoration indicates a late 18th or early 19th century date. The less efficient cooling system (whereby the water merely overflowed through an outlet rather than passing round the condenser tube in a direction counter to the distillate flow) suggests an earlier date than the glass example. A similar arrangement showing a tube passing through a rectangular trough of water is illustrated in a paper, 'Account of a simple Improvement in the Common Still' published in 1808, though there is no provision for flow of cooling water and the scale is rather larger than would be used in a laboratory.[18] This Playfair Collection condenser is unlike any other which survives and may date from between 1790 and 1830.

NOTES AND REFERENCES

1. C E Weigel *Observationes Chemiae et Mineralogicae* (Gottingen 1771).

2. C E Weigel Observationes Chemicae et Mineralogicae ... Pars Secunda (Greifswald 1773). Both volumes of Weigel's thesis were available in translation shortly after their initial publication: J T Pyl *Chemisch-Mineralogische Beobachtungen 1* and *2* (Breslau 1779).

3. *Nouvelle Construction d'Alambic poure faire toute Sorte de Distillation en grand ...* ([Paris] 1781). The copy from the Biblioteca Mediceo-Lorenese, now in the Museo di Storia della Scienza in Florence, bears the inscription in MS 'par J H de Magellan' on the title page as well as further annotations by the author in the text. The title page also mentions that the first edition (presumably this one) is to be distributed gratis in the French provinces, while the second edition (not published?) was to be sold for the benefit of hospitals.

4. J Gadolin 'Beskrifning pa en forbattrad Afkylnings-Anstalt vid Branvins-Brannerier' *Kongl Vetenskaps Academiens Nya Handingar 12* 193 (1791).

5. J Gadolin 'Beschreibung einer verbesserten Abkünlungsanstalt bey Brannteweinbrennereyen' Crell's *Chemische Annalen 1* 369 (1792).

6. J Gadolin 'Forslag at forbattra Brannerie-slangen' *Kongl Wetenskaps Academiens Handingar 39* 283 (1778).

7. Max Speter 'Geschichte der Erfindung des Liebigischen Kühlapparates' *Chemiker Zeitung 32* 3 (1908).

8. Justus Liebig 'Ueber die Verbindungen, welche durch die Einwirkung des Chlors auf Alkohol, Aether, olbindenes Gas und Essiggeist entstehen' *Annalen der Pharmacie 1* 182 (1832), plate 1 ("ich habe eine Zeichnung davon beigelegt, die ich nicht naher beschreibe").

9. J Liebig, J Poggendorff and F Wohler *Handworterbuch der reinen und angewandten Chemie 2* (Brunswick 1842) 531 ("Liebig hat einer Apparat beschrieben, welcher selbst bei raschen Destillationen sehr fluchtiger Flussigkeiten vollkommen abkühlt").

10. Thomas Graham *Elements of Chemistry* (London 1842). On the other hand, Samuel Gray had correctly ascribed the invention to Weigel in his *Operative Chemist* (London 1828), mentioning that one Poissonier had adapted the apparatus for the distillation of spirits in 1779.

11. Neither is the condenser described in *An Explanatory Dictionary of The Apparatus and Instruments of Chemistry* (London 1824) (which has an extensive section on distillation); nor in Andrew Duncan (ed) *Edinburgh New Dispensatory* 11th ed (Edinburgh 1826).

12. D B Reid *Elements of Chemistry* 3rd ed (Edinburgh 1839) 787 ("it is very effectual in condensing completely any vapours that may be led into [it]").

13. J J Griffin *Chemical Recreations* 8th ed (Glasgow 1838) 201.

14. J J Griffin *Chemical Handicraft: A Classified and Descriptive Catalogue* (London 1866) 205. The dimensions given (inner tube 36 inches long, ½ inch diameter; outer tube 24 inches long, 1 inch diameter) are slightly longer and narrower than the Playfair Collection example (object 25).

15. *A Descriptive Catalogue of Chemical Apparatus, Berlin and Dresden Porcelain, Chemical Tests &c. Manufactured and Sold by Charles Button* (London 1846). The condenser was advertised as "Liebig condenser (Brass) German make, with Universal joint and cross joint, so as to elevate to any angle". This was offered at 30 shillings, zinc japanned versions being available at 15-18 shillings.

16. Condensers of this kind are illustrated in Griffin *op cit* (13) 198 and C Greville Williams *A Handbook of Chemical Manipulation* (London 1857) 217. There is an example of this type in the Science Museum, London (Department of Chemistry), inv no 1946-57, made by John Newman, 122 Regent Street, London. It must date from 1827 to 1838 (see E G R Taylor *The Mathematical Practitioners of Hanoverian England* (Cambridge 1966) 400).

17. Those condensers used by Liebig which survive in the Deutsches Museum, Munich, are of this type. See Ernest Child *Tools of the Chemist* (New York 1940) 38.

18. 'Account of a simple Improvement in the Common Still. In a letter from Mr J.Acton' *Nicholson's Journal 21* 358 (1808).

26. Iron bottle
[Iron bottle for making Oxygen. 1858.275.26]

Overall length 470, maximum width 83 mm.

A cast iron bottle, elliptical in cross-section, is fitted at its mouth with a cylindrical iron adaptor into which is jammed a conical iron tube. The bottle contains a powdery white residue which on chemical analysis was shown to have the composition[1]: $CaCO_3$ 47.5% (by weight), $MgCO_3$ 0.16%, Fe_2O_3 56%, Mn negligible.

Figure 38 Iron bottle

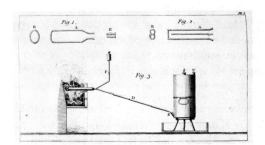

Figure 39 Beddoes' and Watt's portable pneumatic apparatus, from James Watt *Description of a Simplified Apparatus and Portable Apparatus* (Bristol 1796) plate 5

The iron bottle is probably part of an apparatus produced by Thomas Beddoes and James Watt for preparing gases. It was described in a publication of 1796.

Thomas Beddoes (1760-1808) studied medicine under Joseph Black at Edinburgh, graduated MD at Oxford in 1787, and taught as Chemical Reader at Oxford from 1787 to 1792. He advocated the cure of various conditions by the inhalation of gases. From 1793 he lived at Clifton, near Bristol, and in 1798, having received technical help from James Watt (who designed and constructed his apparatus)[2] and financial help from Thomas Wedgwood and others, he opened his Pneumatic Institution. Prior to this, in 1794, he had published *Considerations on the Medical Uses of Factitious Airs*[3], Watt describing the apparatus. A smaller 'portable apparatus' was designed later, and a description of this was published in 1796 as *Supplement . . . Containing a Description of a Simplified Apparatus and of a Portable Apparatus*[4], being advertised as:

> "One oxygene and one hydro-carbonate fire tube, with end pieces, water pipe and cup, conducting pipe, one second-sized air holder and funnel, and an oiled silk bag."

This was offered at a price of £2 12s 6d. The iron bottle in the Playfair Collection closely corresponds with the description of the 'oxygene fire tube' and it is fitted with the 'end piece' and a 'conducting tube' (figure 39 'A', 'E' and 'D'). In operation it was filled with manganese dioxide and heated in a furnace to produce oxygen:

> "The fire tube for oxygene air, is made somewhat like a pocket liquor flask, the flattened form of which permits it to enter between the bars of a common grate. Its dimensions enable it to contain about a pound of manganese, which will generally produce half a cubic foot, or three gallons of air, at one operation."[5]

There was a long correspondence between Joseph Black and James Watt concerning the pneumatic apparatus. The first letter dates from 17 July 1793, when Watt wrote to Black:

> "Doctor Beddoes is applying the antiphlogistic Chemistry to Medicine. Azote and other poisonous airs to cure Consumptions and oxigene for spasmodic asthmas."[6]

A year later Watt wrote to Black to tell him that he had devised a pneumatic apparatus for Beddoes.[7] Black replied asking for a copy of a publication describing it, and further correspondence followed.[8] In 1796, Black accepted the offer of a portable pneumatic apparatus from Watt[9] and after a seven month delay in it being sent he acknowledged its receipt:

> "I had the pleasure to receive your letter of the 1st June and more lately the apparatus perfectly safe. I am much pleased with it".[10]

It seems that Black demonstrated Watt's apparatus at his last session of classes in 1796-7. He wrote to Watt on 13 June 1797, thanking him "I am under great obligations to you for the present of your portable pneumatic apparatus it has been very usefull to us in the chemical course for some of the experiments."[11] The mention of 'us' implies that Thomas Charles Hope was, by this time, helping Black deliver his lectures (see chapter 3). As late as 1820 Hope demonstrated the production of oxygen to his class by heating manganese dioxide in an iron bottle.[12] This method of preparation of the gas was less convenient than by heating potassium chlorate because the temperature required was higher ($530^\circ C$ compared with $350^\circ C$), and it is recorded that it took half an hour of heating for Hope's apparatus to produce oxygen.[13]

That the remaining residue in the bottle is calcium carbonate (mixed with rust) is not easy to explain. One possibility is that it was never used, and this assumption is reinforced by the existence of a layer of black paint on the surface of the bottle. In fact Watt had written in his pamphlet:

> "When a fire tube is used for the first time, any air which is prepared in it has a bad smell: oxygene air in such cases contains a larger portion than usual of fixed air, and the hydrocarbonate of sulphurated hydrogene. These have been with justice imputed to the carbone and sulphur contained in the cast iron of the fire tube ...
> Oxygene air is rendered more pure and more free from fixed air, by preparing the fire tube, by heating it full of quick lime ... then emptying out the quick lime, and filling it with manganese in course powder as usual."[14]

Another possibility is that the bottle was used for the production of carbon dioxide: Hope mentions "To make Carbonic Oxide – Take well dried Chalk ... put into Wedgewoods Iron bottle".[15]

Neither possibility is totally plausible: Black records having used Watt's bottle[11] and it is likely that a significant trace of manganese dioxide would remain even if it was later used for carbon dioxide production. It is feasible that the Playfair Collection bottle is not that supplied by Watt, but one acquired later by Hope (they were offered by a London dealer from at least as early as 1805[16]). Possibly the experiments demonstrated by Black in 1797 did not include the preparation of oxygen by Beddoes' and Watt's portable pneumatic apparatus.

The bottle was exhibited in 1976 at the 'Edinburgh and Medicine' exhibition at the Royal Scottish Museum.[17]

NOTES AND REFERENCES

1. Analysis performed by Dr J H Duncan, Department of Chemistry, University of Glasgow, January 1972.
2. F F Cartwright 'The Association of Thomas Beddoes MD with James Watt FRS' *Notes and Records Royal Soc 22* 131 (1967).
3. Thomas Beddoes and James Watt *Considerations on the Medicinal Use of Factitious Airs and on the Manners of obtaining them in Large Quantities* (Bristol 1794).
4. James Watt *Supplement to the Description of a Pneumatic Apparatus ... added to Consideration on the Medicinal Use of Factitious Airs* (Bristol 1796).
5. *Ibid* 14, figure 1 A, B, E.
6. E Robinson and D McKie *Partners in Science* (London 1970) 195, letter 139.
7. *Ibid* 207, letter 148 (letter Watt to Black, 31 August 1794).

8. *Ibid* 208, letter 149 (letter Black to Watt, 9 September 1794). The publication (ie Beddoes and Watt *op cit* (3)) was sent to Black, but lacking the plates (*ibid* 209, letter 150 (Black to Watt 28 October 1794)). Watt apologised (*ibid* 210, letter 151 (Watt to Black 8 December 1794)) and offered to send Black some 'manganese' (presumably manganese dioxide). Black accepted this offer (*ibid* 212, letter 152 (Black to Watt 13 December 1794)).

9. *Ibid* 219, letter 157 (letter Black to Watt, 4 January 1796 "I long to see your improvements of the apparatus — the portable one will be a most acceptable present to me"). Watt replied to Black on 7 January saying that he had hoped to have had the apparatus sent already but that there were delays (*ibid* 220, letter 158). The apparatus had still not been sent by 1 June, when Watt wrote to Black saying that it was being sent that day (*ibid* 223, letter 159).

10. *Ibid* 225, letter 160.

11. *Ibid* 277, letter 184.

12. Edinburgh University Library MS Gen 270 'List of Specimens' ("To procure oxygene gas . . . iron vessel charged with ... Manganese in course powder . . . Gas generally comes in ½ hour").

13. The thermal decomposition of potassium chlorate for the production of oxygen had been suggested by C L Berthollet in 1788 (*Obs sur la Physique 33* 222 (1788)) but it required special preparation of the chlorate. The mineral pyrolusite (manganese dioxide) was relatively easily obtainable.

14. Watt *op cit* (4) 20.

15. Edinburgh University Library MS *op cit* (12).

16. Frederick Accum *Catalogue of Chemical Preparations, Apparatus and Instruments* (London 1805) 18 ("Cast-iron Retorts, for the production of oxygen and other gases by the culinary fire, 12s"). A bottle very similar to Watt's was illustrated in Accum's *System of Theoretical and Practical Chemistry* 2nd ed. (London 1807) plate VII figure 12.

17. R G W Anderson and A D C Simpson *Edinburgh and Medicine* (Edinburgh 1976) 45, item 263.

27. Pattern for beehive shelves (?)
[Model of Stool for Collecting gases in a pneumatic trough. 1858.275.27]

Base diameter 121, height 96mm.

Wooden (pear) dome with rim, constructed from two turned pieces of wood. A semicircular hole is cut in the side of the lower piece and a small circular hole is cut in the apex.

Figure 40   Pattern for beehive shelves

The beehive shelf was developed to support water-filled gas jars in pneumatic troughs while they were being filled with gas bubbling from tubes beneath them. It evolved from what was literally a shelf built on to a pneumatic trough. The first to be illustrated (in 1765) was that of the physician William Brownrigg who supported inverted water-filled bottles on a wooden board.[1] Joseph Priestley first rested his gas jars on thin flat stones, but later his pneumatic trough had a built-in wooden shelf.[2]

The earliest form of the beehive shelf was an earthenware cube with a smaller cube removed from one edge. This was illustrated by the Glasgow instrument maker and dealer John Joseph Griffin in his *Chemical Recreations* of 1838.[3] The later, and more usual form, was described in the same volume:

> "The shelf is an inverted stone pan, cylindrical on the outside, but shaped like a bee hive within. It is 4 inches broad, and 3½ inches high. On the side it has a round opening of 2 inches diameter, to admit the entrance of retort necks and delivering tubes, and in the top it has an opening half an inch in diameter, to permit the passage of gas from the beehive below into the jars placed upon it."[4]

Griffin later claimed that he had invented the beehive shelf.[5]

The instrument in the Playfair Collection could not have been used for collecting gases: it is made of wood (and would float in a trough of water) and it has a rounded top. It may have been constructed to demonstrate the principle of the apparatus to the chemistry class. More likely, it may have been used as a mould (in that sense, a 'model') round which clay would be packed in the production of beehive shelves.

A ceramic beehive shelf of usual design, impressed 'KEMP & Co/ EDINBURGH' (which must date from between 1835 and 1887), survives in the Chemistry Department, University of Edinburgh.

NOTES AND REFERENCES

1. William Brownrigg 'An Experimental Enquiry into the Mineral Elastic Spirit . . .' *Phil Trans Royal Soc 55* 218 (1765).

2. Joseph Priestley *Experiments and Observations on Different Kinds of Air 1* (London 1774) 7. See also Lawrence Badash 'Joseph Priestley's Apparatus for Pneumatic Chemistry' *J Hist Med 19* 139 (1964).

3. J J Griffin *Chemical Recreations* (Glasgow 1838) 212. For details of Griffin's activities see Playfair Collection object 19, reference 4. Partington thought that the beehive shelf was first depicted in Thomas Graham *Elements of Chemistry* (London 1842) 247 (see J R Partington *A History of Chemistry 3* (London 1962) 250). C Greville Williams in his *A Handbook of Chemical Manipulation* (London 1857) 297 correctly ascribed the invention to Griffin ("By far the most convenient . . . method is termed a bee-hive shelf. The last-named piece of apparatus consists of a vessel shaped like a bee-hive, only flat at the top, so as to support a gas-jar . . . The use of this little contrivance (the invention, it is believed, of Mr. J.J. Griffin)").

4. Griffin *ibid* 215.

5. J J Griffin *Chemical Handicraft* (London 1866) 227.

28, 29. Two concave reflectors
[A pair of brass reflectors, 12½ ins diameter. 1858.275.28]

Diameter (of both) 315, depth 62mm.

Hammered brass reflectors of parabolic shape with slight rim at the edge. One (28) has a bright polished finish, the other (29) has a blackened matt surface.

30. Concave reflector
[A tin reflector, 12¼ ins diameter. 1858.275.29]

Diameter 313, depth 56mm.

Cast brass reflector (with a roughened (solder?) surface); at the back is attached a short length of cylindrical tube fitted with a round headed screw to allow fixing to a vertical stand (missing).

Figure 41 Concave reflectors (Object 28, left; object 29, right; object 30 foreground

Experiments on the focussing of radiant heat by the use of concave mirrors were first undertaken in the second half of the 17th century. The activities of the Accademia del Cimento of Florence have been mentioned in connection with another object in the Playfair Collection (wire cage, object 6). Thomas Thomson wrote a brief history on the reflection of heat by mirrors in 1830.[1] He said that Edme Mariotte performed "the first crude attempt to examine radiation" in 1682 at a meeting of the Academie Royale des Sciences, finding that "the heat of a fire reflected in a burning mirror is sensible in its focus".[2] J A Nollet described an experiment using two parallel concave mirrors in 1754 which interested Joseph Black.[3] In February 1768 the latter wrote to James Watt:

> "I am persuaded that cokes act mostly by a radiation like that of the sun. There is a pretty experiment in Nollets Lecons de Physique; — he set two mirrors, (made of pasteboard gilt), parallel to one another, and face to face, in the opposite sides of a room; in the focus of one, a bit of charcoal, and in that of the other, a little gunpowder; — he blew upon the charcoal to brighten it, and the gunpowder took fire."[4]

H B de Saussure detected heat radiated from a hot bullet in the focus of one mirror in the focus of another[5] and in 1790 M A Pictet published an account in which he described his mirrors in some detail and he recorded that he detected a rise in temperature of ten degrees using a mercury thermometer heated by a hot (though not luminous) bullet.[6]

Experiments were also undertaken in Edinburgh. In 1804, the professor of natural philosophy, John Leslie, published his *Experimental Inquiry into the Nature and Propogation of Heat,* in which he described a long series of experiments on radiant heat using differential thermometers and pairs of parallel concave metal mirrors. (Leslie's differential thermometers have already been described: see Playfair Collection, objects 10 and 11). In his description, he makes the interesting point that it was difficult to find a craftsman capable of making the mirrors:

> "The principal articles of the apparatus were *specula* or reflectors made of tinned iron. Of these I had several, of different dimensions; from twelve to about fourteen inches in diameter, and with a concavity from 1¾ to near 2½ inches [45-62mm]. It cost me no small trouble to obtain what I wanted. I had to make repeated trials before I could find an artist skilful enough to execute the reflectors with any tolerable precision."[7]

Metal mirrors were produced commercially shortly afterwards: the dealer Frederick Accum advertised mirrors on stands for between £2 15s 0d and £10 10s 0d in 1817.[8]

The mirrors in the Playfair Collection lack the means of support for experimental use. In two cases (objects 28 and 29) they would be supported in a frame holding the rim, as described by Leslie:

> "When a reflector was used, it was supported from the table in an upright position by the help of a small wooden frame or stand, consisting of two narrow perpendicular pieces, extending somewhat above the centre of the reflector but rather less distant asunder than its diameter, and morticed into a pretty broad piece with cross claws at the ends. On the inside, towards the top of each of those pillars, and near the middle of the flat piece, a slight groove was cut, through which the reflector was let down; and though it was held firm by the gentle spring of the wood, it could easily be removed at pleasure without deranging the stand."[9]

In the case of the third (object 30) a stand (similar to a retort stand) would be required. Such an arrangement was referred to in a dictionary of apparatus of 1824:

> "These concave reflectors are usually made of planished tinned iron plate, about 12 inches diameter. . . they must be fastened by sliding rings and screws on metallic stands."[10]

A third method of support (of which there is no example in the Playfair Collection) was by vertical suspension from three points on the circumference of the mirror.[11]

The Playfair Collection mirrors almost certainly date from the period of Thomas Charles Hope's teaching. Hope mentioned the use of such apparatus in his notes from which he lectured. In an evelope of notes marked "Ignition" he wrote: "Put to test, try the reflection by Black Speculum on Mur of Silver".[12] In his 'List of Specimens' (which Hope had made out for his assistant to prepare for lecture demonstrations) are the instructions: "Brass Mirrors (N.B. Oxalic Acid in Satd Solution effectual in brightening)".[13] The mirrors are made of brass rather than tin or tinned iron which is the most frequently mentioned material. The surface of one of the mirrors (object 30) has been made deliberately uneven by pouring a molten solder (?) over it, perhaps to reduce its power of reflection.

NOTES AND REFERENCES

1. Thomas Thomson *Heat and Electricity* 2nd ed (London 1840) 115.
2. Mariotte's experiment is mentioned in *Histoire de l'Académie Royale des Sciences 1* 344 (1733).
3. [J A] (Abbé) Nollet *Leçons de Physique Experimentale 5* (Amsterdam 1756) 208.
4. E Robinson and D McKie *Partners in Science* (London 1970) 10, letter (Joseph Black to James Watt, 19 February 1768).
5. H B de Saussure *Voyages dans les Alpes 2* (Geneva 1786) 353.

6. M A Pictet *An Essay on Fire* (London 1791) 86 (Translation of M A Pictet *Essais de Physique 1* (Geneva 1790)). ("We placed two concave mirrors of tin... opposite each other in a large room, at the distance of 12 feet 2 inches. These mirrors are a foot in diameter, and their convexity that of a sphere whose radius is 9 inches, and they are but moderately polished.") Pictet also used air thermometers because "thermometers of air are known to possess a supreme degree of sensibility" (*ibid* 95).

7. John Leslie *An Experimental Inquiry into the Nature and Propogation of Heat* (London 1804) 2.

8. Frederick Accum *Descriptive Catalogue of the Instruments employed in Chemistry* (London 1817) 11 ("Concave metallic mirrors, with stands, for rendering sensible the reflection of radiant heat, unaccompanied by light").

9. Leslie *op cit* (7) 5.

10. *An Experimental Dictionary of the Apparatus and Instruments of Chemistry* (London 1824) 263, plate 14 fig 10.

11. W T Brande *A Manual of Chemistry* (London 1819) 65.

12. Edinburgh University Library MS Gen 268, envelope 27. The quotation may refer to an experiment to demonstrate the decrepitation of silver sulphate at 300°C, and perhaps its melting to a brownish liquid at 660°C. The "Black Speculum" refers to its surface, not a previous owner.

13. Edinburgh University Library, MS Gen 270 'List of Specimens'.

31. Thaumaturgical device
[Tetragrammaton. 1858.275.31]

Sides of triangle 140, 140, 134, approximate diameter of sphere 38 (1.5 inches), sides of box 200 mm.

A copper near-equilateral triangle with a small rectangular projection (pierced with a hole) from the smallest side (for suspension?) is mounted with a slightly flattened sphere of glass through a hole in the centre. Engraved on one side are two concentric circles surrounding the crystal which contain the word 'TETRAGRAMMATON' and the symbols of a pentangle, a Shield of David containing the Hebrew letter *lamel* and a Maltese cross. On the reverse side is engraved two concentric circles containing the names of the archangels: 'Gabriel', 'Raphael', 'Uriel' and 'Michael'. Outside the circles appears a Shield of David containing the Hebrew letter *lamel*.

Figure 42 Thaumaturgical device

This is the only item from the Playfair Collection which has no connection with conventional laboratory operations. It is almost certain that it must have been the property of William Gregory. Gregory took an active interest in mesmerism, animal magnetism and phrenology, and had the reputation of being gullible. He participated in public displays of pseudo-scientific rites (at which it is possible that this device was used). Gregory translated Carl von Reichenbach's work on animal magnetism and wrote works of his own on this subject and on mesmerism. These aspects of his character are dealt with at greater length in Chapter 4.

The device may contain the 'magic crystal' referred to by Gregory in his *Letters to a Candid Inquirer, on Animal Magnetism* of 1851: "I have been informed of two other magic crystals... A fourth is in my possession and I hope to obtain its history."[1] In the same work he wrote of the use to which glass speres were currently being put:

"I have been informed, on good authority, that round or oval masses of glass are made in England, and sold at a high price, to the ignorant, for the purpose of divination... It is probably a process of magnetisation, as water is magnetised. The purchaser is then directed to gaze into the crystal, concentrating her thoughts on the person she wishes to see... Now, I believe, that by the gazing and concentrating of the thoughts, aided by the odylic influence of the glass, she may be rendered more or less lucid and thus see or dream of the absent person. So that dealers in these crystals are not mere imposters, but, as I suppose, trade in a natural truth, imperfectly known to themselves."[2]

The thaumaturgical device in the Playfair Collection, which presumably can be associated with Gregory's crystal-gazing activities, bears a very close resemblance to an instrument described by Francis Barrett in 1801. Barrett's work *The Magus, or Celestial Intelligencer* contains translations of the writings of Johannes Trithemius who evidently discussed the device and its use. The description by Barrett is as follows:

"Of the making of the CRYSTAL and the Form of Preparation for a VISION PROCURE *of a lapidary a good clear pellucid crystal, of the bigness of a small orange, i.e. about one inch and a half in diameter; let it be globular or round each way alike; then, when you have got this crystal, fair and clear, without any clouds or specks, get a small plate of pure gold to encompass the crystal round one half; let this be fitted on an ivory or ebony pedestal, as you may see more fully described in the drawing, (see the Plate, fig 1). Let there be engraved a circle (A) round the crystal with these characters around inside the circle next the crystal . . . afterwards the name "Tetragrammaton". On the other side of the plate let there be engraven "Michael, Gabriel, Uriel, Raphael"; which are the four principal angels ruling over the Sun, Moon, Venus and Mercury."*[3]

NOTES AND REFERENCES

1. William Gregory *Letters to a Candid Inquirer, on Animal Magnetism* (Edinburgh 1851) 313.
2. *Ibid* 315.
3. Francis Barrett *The Magus, or Celestial Intelligencer* (London 1801) Part IV 135. A similar description and discussion of the method of using such devices appears in G F Kunz *The Curious Lore of Stones* (London 1913) 181.

32-35. Four crucibles
[Iron Crucibles (3 specimens). 1858.275.32]

Base diameter (of all specimens) 45, height 67 mm.

Four closely similar iron crucibles with black paint on inside and outside. The original catalogue entry mentions only three, though all four appear to form part of the collection, bearing the same registration number.

Figure 43 Iron crucibles

Iron crucibles have never been used extensively in chemical laboratories, those of ceramic material being most favoured. A writer explained in 1824:

> "Iron crucibles resist heat extremely well; but the air aided by the action of the fire, oxidizes them very speedily, and saline matters readily act upon them; even earths become coloured in iron crucibles, so that they cannot be employed in fusion, except in very few cases."[1]

In his *Chemical Manipulation* of 1827, Michael Faraday put forward a similar view:

> "An iron crucible is occasionally required; though but seldom for experiments of ignition. On the whole, though a more fusible material, it is better of cast than of wrought metal."[2]

J J Griffin's first extensive catalogue, that of 1838, makes no mention of iron crucibles, but by 1866 both cast iron and wrought iron crucibles were being offered by the firm of Richard Griffin and Company.[3]

Joseph Black discussed crucibles in his lectures, though he made no mention of those made from iron:

> "The materials of crucibles are always earthy compositions, made with great care, as they are often exposed to great heats. There are but two varieties in common use — the Hessian, and the Ipsian, or Austrian, or black lead."[4]

It seems most probable that the examples in the Playfair Collection date from the second quarter of the 19th century.

NOTES AND REFERENCES

1. *An Explanatory Dictionary of the Apparatus and Instruments of Chemistry by a Practical Chemist* (London 1824) 84.
2. Michael Faraday *Chemical Manipulation* (London 1827) 286.
3. J J Griffin *Chemical Recreations* 8th ed (Glasgow 1838); J J Griffin *Chemical Handicraft* (London 1866) 142.
4. Joseph Black *Lectures on the Elements of Chemistry . . . Published by John Robison 1* (Edinburgh 1803) 294.

36. Series of absorption bottles
[Arrangement for saturating water with Ammonia, hyxdrochloric (sic) acid &c. 1858.275.33]

Base 1168 x 228, diameter of bottles (approximately) 140, height of bottles 230 mm.

A wooden base (mahogany) with a shelf at one end and carved carrying handles supports four pairs of vertical iron bars between each pair of which is held a clear glass bottle. All four bottles have a ground glass neck at the top, the stoppers to three of which survive. All bottles have two additional tubulatures; in three cases these are just below the shoulders on opposite sides of the bottles, the fourth has one tubulature in the shoulder itself (containing a ground glass stopper to which has been attached a paper label with the manuscript inscription "Neck Stopper of one of Black's Bottles") and one just below the shoulder. Three of the bottles are joined between their tubulatures with a lute.

Figure 44 Series of absorption bottles

This piece of apparatus would seem to have been designed to perform a similar function to a series of Woulfe bottles, trapping the constituents of a mixture of gases by dissolving them in solutions. The apparatus devised by Peter Woulfe, described in 1768, was an improved distillation apparatus with a trap for condensing volatile products.[1] The so-called Woulfe bottle was not described by Woulfe at all. It may have been introduced in Paris in about 1773, possibly by J B M Bucquet.[2] A L Lavoisier illustrated two and three-neck Woulfe bottles in his *Traité Elémentaire de Chimie* of 1789.[3] The three-necked bottles are described in some detail. Lavoisier mentioned that J H Hassenfratz had first proposed using a third neck to prevent water being sucked into the apparatus from the pneumatic trough.[4]

Woulfe bottles have retained the design of Lavoisier and Hassenfratz ever since. This might be taken to imply that the Playfair Collection bottles pre-date these. This, however, cannot be assumed. Series of absorption bottles more similar to the Playfair Collection examples were illustrated in 1805[5] and 1830.[6] Woulfe bottles on a wooden tray were advertised by the firm of Richard and George Knight in 1811.[7] The Playfair Collection bottles were probably mounted on the base to allow them to be carried into the chemistry class for demonstrations. The apparatus is rather crude, the bottles probably originating as pharmacist's bottles.[8] They have been clumsily adapted by the addition of tubular openings just below the shoulders. The bottles were originally joined by luting (a substance used for joining glass apparatus consisting of, for example, clay, sand, flour, linseed oil, etc.)[9]; this remains around the glass tubulatures.

It is not clear why the label on one of the ground glass stoppers implies that the bottle might be associated with Black. If the apparatus were constructed by or for him, it is unlikely to have been made prior to 1785, for on 27 March of that year James Lind wrote a letter to Black describing how to grind a stopper into a glass neck.[10]

Evidently this was in reply to an enquiry made by Black; Lind had written to Tiberius Cavallo for information. It is by no means easy to date this apparatus: the bottles could date from the last years of Black's work or from any part of Hope's period of teaching.

A very similar bottle in pale green glass survives in the Pharmaziehistorisches Museum, Basel, though the tubulatures are not ground.[11]

NOTES AND REFERENCES

1. P Woulfe 'Experiments on the Distillation of Acids, volatile Alkalies, &c Shewing how they may be Condensed, and how thereby we may Avoid Disagreeable and Noxious Fumes' *Phil Trans Royal Soc 57* 517 (1768).

2. J R Partington *A History of Chemistry 3* (London 1962) 301.

3. A L Lavoisier *Traité Elémentaire de Chimie* (Paris 1789) plate 4 figures 1, 13, 14, 15.

4. A L Lavoisier *Elements of Chemistry . . . translated by Robert Kerr* (Edinburgh 1790) 394 ("In experiments of this kind, I for a long time met with an almost insurmountable difficulty, which must have obliged me to desist altogether, but for a very simple method of avoiding it, pointed out to me by Mr Hassenfratz. The smallest diminution in the heat of the furnace . . . cause frequent reabsorptions of gas; the water in the cistern of the pneumato-chemical apparatus rushes into the last bottle. . . This accident is prevented by using bottles having three necks, into one of which, in each bottle, a capillary glass tube is adapted.")

5. 'Improved Construction of Woulfe's Apparatus. By WB' *Nicholson's Journal 10* 180 (1805).

6. Arthur L Porter *Chemistry of the Arts 1* (Philadelphia 1830) 203, figure 89 ("A very convenient distillatory apparatus has been invented by Dr De Butt, of Baltimore" (probably Elisha de Butts)).

7. *Catalogue of Apparatus and Instruments for Philosophical, Experimental, and Commercial Chemistry, Manufactured and Sold by R & G Knight* (London 1811) 19 ("Woulf's Apparatus, consisting of Three Bottles, with Three Necks each, mounted with Conducting and Safety Tubes, in a Wooden Tray, complete, to contain Half a Pint, 18s").

8. Similar shaped bottles (though with single necks) are described in J K Crellin and J R Scott *Glass and British Pharmacy 1600-1900* (London 1972) 42, item 64, figure 53.

9. Lutes were discussed by Black; see Joseph Black *Lectures on the Elements of Chemistry . . . Published by John Robison 1* (Edinburgh 1803) 335. Though rubber tubing for joining apparatus may date from the mid 18th century, it was by no means used widely for another century, see Partington *op cit* (2) 190 note 2, and L Badash 'Joseph Priestley's Apparatus for Pneumatic Chemistry' *J Hist Med 19* 147 (1964).

10. Edinburgh University Library MS Gen 874 vol 4, f11. Black suggested to James Watt that he use glass or metal stoppers instead of corks on his pneumatic apparatus, though this was rejected because of the expense (see E Robinson and D McKie *Partners in Science* (London 1970) 225 (letter Black to Watt, 28 July 1796) and 228 (letter Watt to Black 9 October 1796)). Black also suggested the use of glass stoppers for phials containing deliquescent or corrosive substances (see Black *op cit* (9) 300).

11. Pharmaziehistorisches Museum, Basel *Die Sammlung. Darstellung alter Arztinstrumente, Apotheker-Gefässe...3* (Basel 1974) 60, figure 39.

37. Bottle with a tap
[Apparatus for transferring gases to a bladder. 1858.275.34]

Overall height 263, diameter 96 mm.

A clear glass bottle which has an unground pontil mark is fitted with a brass tap (to which is joined a right angled adaptor) and is sealed through the neck by means of a cork, made air-tight with black wax.

Figure 45 Bottle with a tap

The exact function of this apparatus may never be ascertained with certainty. It is crudely constructed and may have been made in the University chemistry laboratory to serve a very specific function. It is not, as originally described in the Museum Register, an apparatus for transferring gases to a bladder. Such apparatus was superficially similar, consisting of a wide glass tube with a tap at the top, the bladder being attached to the tap. The bladder was filled by lowering the apparatus in a trough of water, water pressure forcing the gas upwards. The fundamental difference is that such an apparatus had no base so that the gas could be collected over water prior to filling the bladder.[1]

129

It is not easy to speculate how this bottle may have been used. It is too heavy to be used for weighing a gas to determine its specific gravity. It is not suitable for heating and it is not strong enough to be used as an eudiometer (for exploding gases to investigate their reactions). A most likely function might be in some fairly gentle process carried out at room temperature for which only one inlet or outlet was required. A possible experiment could be the demonstration of the high solubility of gases such as ammonia or hydrogen chloride: the bottle would be evacuated of air, filled with gas and then inverted over a trough of water, which would lead to the water being sucked into the bottle. It might also be used to generate a gas. Joseph Priestley used a similar bottle in his experiments on the impregnation of water with carbon dioxide, partly filling a bottle with chalk over which water was poured. When the gas was required a little sulphuric acid was added. Priestley did not use a bottle with a tap, though it would have facilitated the attachment of the bladder had he done so.[2] A.L. Lavoisier used a bottle with a tap to collect the froth spilling over from a fermentation.[3] To associate the bottle with these, or any other experimental use, with any confidence, would be unjustifiable.

NOTES AND REFERENCES

1. This standard piece of pneumatic apparatus was widely described. See, for example, *An Explanatory Dictionary of the Apparatus and Instruments of . . . Chemistry* (London 1824) 251, or J J Griffin *Chemical Recreations* 8th ed (Glasgow 1838) 213.

2. Joseph Priestley *Experiments and Observations on Different Kinds of Air* 2nd ed (London 1776) 280, plate 2 (bottle labelled 'e').

3. A L Lavoisier *Elements of Chemistry . . . translated by Robert Kerr* (Edinburgh 1790), 402, plate 10 fig. 1 (bottle labelled C).

38. Graduated glass vessel
[Graduated glass receiver. 1858.275.35]

Overall length 322, diameter at midpoint of glass tube 47 mm.

A glass tube of slightly conical shape is drawn into necks at each end. Brass tubes with brass taps (neither signed) are luted to the necks and terminate in screw threads. The tube has been graduated, by scratching with a sharp point, into six numbered divisions, each subdivided into half units. The volumes corresponding to the five divisions between 1 and 6 are found to be 1.03±0.04 cubic inches (16.9±0.7 ml).

Figure 46 Graduated glass vessel

The exact way in which this apparatus was used is not certain. A graduated tube fitted at either end with stopcocks could conceivably be used for a variety of purposes, but it is likely to have been used to measure small volumes of gases and its most probable function is that it was used to make quantitative measurements in the absorption by water of a moderately soluble gas.

Experiments to determine the solubility of carbon dioxide in water had started soon after Black had identified the gas as being distinguishable from atmospheric air. In 1766, Cavendish experimented with a tube closed at one end. The apparatus was described, but not illustrated:

> "In order to find how much fixed air water would absorb, the following experiment was made. A cylindrical glass, with divisions marked on its sides with a diamond, shewing the quantity of water which it required to fill it up to those marks, was filled with quicksilver, and inverted into a glass filled with the same fluid. Some fixed air was then forced into this glass, in the same manner that it was into the inverted bottles of water . . ."[1]

A F de Fourcroy performed similar experiments.[2] In 1803, William Henry published his experiments on a number of gases, among which were the moderately soluble carbon dioxide, hydrogen sulphide and nitrous oxide.[3] Henry used two kinds of tube for his absorptions: a cylindrical one with a brass stopcock at the top, with a manometer branching off at the other end and a second brass stopcock beneath it; and a pear shaped one, with a brass stopcock at the top, and a ground glass stopper at the base. Both types were engraved with graduations. The Playfair Collection is very similar to the first type of tube, but without provision for a manometer. The experiment would still be possible, however, using the apparatus in the manner of the latter. Henry had attended Joseph Black's lectures in 1795-96, and had worked with Thomas Thomson in Edinburgh in Thomson's private laboratory, though his research on the solubility of gases was conducted at Manchester.[4] He mentioned in his paper that his graduated tube was four inches in length and two in diameter. The Playfair Collection tube has dimensions of four inches by one and a half. Unfortunately the taps do not have a manufacturer's name stamped on them, which was a fairly common practice. The production of taps with square thumb pieces was later than those with round ended ones, and indicates a 19th rather than an 18th century date.

NOTES AND REFERENCES

1. Henry Cavendish 'Experiments on Factitious Airs' *Phil Trans Royal Soc 56* 159 (1767).
2. A F de Fourcroy *Système de Connaissances Chimiques 1* (Paris 1801) 215 (quarto issue).
3. William Henry 'Experiments on the Quantity of Gases absorbed by Water' *Phil Trans Royal Soc 93* 29 (1803) plate 1.
4. David F Larder 'Thomas Thomson's Activities in Edinburgh 1791-1811' *Notes and Records Royal Soc 24* 295 (1969); W V Farrer, Kathleen R Farrer and E L Scott 'The Henrys of Manchester Part 3. William Henry and John Dalton' *Ambix 21* 208 (1974).

39. Stoppered glass bottle
[Bottle supposed to have been used by Dr Joseph Black in preparing Carbonic Acid. 1858.275.36]

Overall height 330, maximum diameter 180 mm.

A clear glass pear-shaped bottle with a prominent rim is sealed with a ground glass stopper with a square-shaped thumb piece; the pontil mark remains at the base. The bottle contains a white powdery deposit which on chemical analysis was shown to have the composition: $CaCO_3$ 91% (by weight), phosphate (expressed as P_2O_5) 3.7%, $MgCO_3$ 0.56%, Fe_2O_3 trace.[1]

Figure 47   Stoppered glass bottle

This object can be treated separately from the other chemical glassware. (Playfair Collection, objects 40-63) as it is made from clear glass rather than from green glass. Its shape is similar to a set of green pharmacy carboys in the Wellcome Collection which are of intermediate form between globe-shaped and pear-shaped examples and it is possibly of the late 18th century.[2] It has been suggested that this bottle may be of later date than the green glass, though there is no good reason why this should be so.[3]

There are no valid grounds for believing that the bottle was used by Black, as stated in the original Museum Register entry. He did not describe the vessels he used in his experiments in which he prepared carbonic acid (a solution of carbon dioxide in water) or carbon dioxide gas.[4] The calcium carbonate deposit might suggest that it was used for the generation of carbon dioxide on the addition of a mineral acid.

The square shaped thumb piece is similar to those on the absorption bottler (Playfair Collection, object 36). Black's references to ground glass stoppers are mentioned in the entry for that apparatus.

NOTES AND REFERENCES

1. Analysis performed by Dr H J Duncan, Department of Chemistry, University of Glasgow, January 1972.
2. J K Crellin and J R Scott *Glass and British Pharmacy 1600-1900* (London 1972) 6, 35 (items 48a-e, fig. 38).
3. Revel Oddy 'Some Chemical Apparatus blown by hand in the late 18th to early 19th century' *Annales du 5e Congrès de l'Association Internationale pour l'Histoire du Verre* (Liege 1972) 231.
4. Joseph Black *Lectures on the Elements of Chemistry . . . published by John Robison 2* (Edinburgh 1803) 99. 'Carbonic Acid' may refer to the gas — it was not a term used rigidly.

40-63. Chemical glassware

Prior to listing the remaining glassware, the Museum Register records:

"No. 38 to No. 45 are various forms of glass vessels used by Dr Joseph Black between 1766 & 1799. They are supposed to have been made at Leith".

40-42 Three cucurbits

[Solution Glasses figured in Dr Black's Lectures (3 specimens). 1858.275.38]

(40) height 335, maximum diameter 105; (41) height 320, maximum diameter 130; (42) height 360, maximum diameter 130 mm.

Gourd-shaped flasks of dark green glass containing bubbles, with lips. All have unground pontil marks.

Figure 48 Cucurbit (object 40) and alembic (object 56) (left), cucurbits (objects 41 and 42), centre and right respectively)

135

43, 44 Two globular bottles
[Flasks or receivers of various forms. (10 specimens). 1858.275.39]

(43) height 230, diameter 200; (44) height 235, diameter 205 mm.

Round flasks of thick dark-green glass. The necks are broken in each case close to the body.

45, 46 Two ellipsoidal flasks
[Flasks or receivers of various forms. (10 specimens). 1858.275.39]

(45) height 595, maximum diameter 170; (46) height 610, maximum diameter 185 mm.

Large tall-necked, round bottomed pale-green glass flasks. (45) has a broken neck, (46) has a splayed-out rim.

47. Straight-sided flask
[Flasks or receivers of various forms. (10 specimens). 1858.275.39]

Height 220, diameter 85 mm.

Round bottomed flask with steeply sloping shoulders, broken neck, in pale-green glass.

Figure 49   Globular bottles (objects 43 and 44, left and centre respectively), ellipsoidal flasks (objects 45 and 46, second and fourth from left respectively), straight sided flask (object 47, right)

48, 49 Two ovoidal flasks

[Flasks or receivers of various forms. (10 specimens). 1858.275.39]

(48) height 200, maximum diameter 115; (49) height 340, maximum diameter 170 mm.

Flasks in thick dark-green glass with broken necks; (48) has pronounced shoulders, (49) has steeply sloping shoulders.

50. Florentine flask

[Flasks or receivers of various forms. (10 specimens). 1858.275.39]

height 695, maximum diameter 170 mm.

Round bottomed flask in dark green glass with very long neck of uneven diameter (maximum width is near the top).

51. Flask with constricted neck

[Flasks or receivers of various forms. (10 specimens). 1858.275.39]

Height 450, maximum diameter of lower portion 80 mm.

Round-bottomed vessel of dark-green thick glass in two portions divided by a constriction, the upper neck broken off.

Figure 50 Ovoidal flasks (objects 48 and 49, left and third from left respectively), Florentine flask (object 50, second from left), flask with constricted neck (object 51, right)

52-54. Three retorts
[Retorts (3 specimens). 1858.275.40]

(52) overall height 355, length 505, diameter of bulb 200; (53) overall height 105, length 445, diameter of bulb 85; (54) overall height 220, length 280, diameter of bulb 125 mm.

Retorts made in thick dark-green glass, with broken necks. (52) and (54) are inexpertly made, there being constrictions where the neck has been bent over in manufacture.

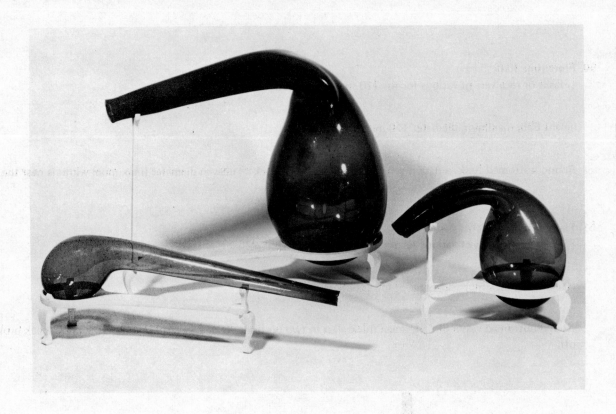

Figure 51 Retorts (object 52, centre; object 53, left; object 54, right)

55. Cucurbit
Lower portion of an Alembic, called a cucurbit by Dr Black. 1858.275.41]

Height 340, maximum diameter 135 mm.

Round bottomed flask in thin pale-green glass, with carefully made rim.

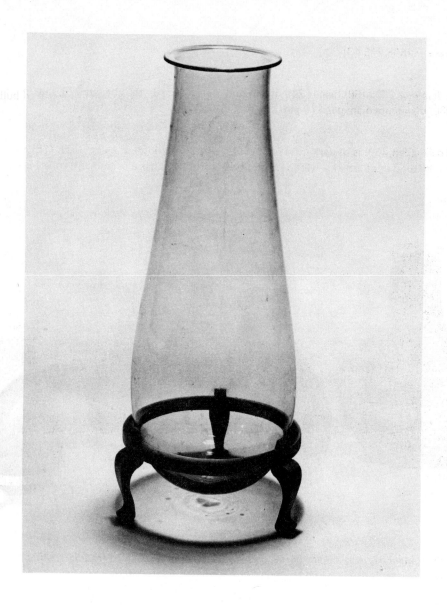

Figure 52 Cucurbit

56. Alembic
[Top or Capital of an Alembic. 1858.275.42]

Overall height of dome 190, diameter of dome 190, length of spout from lowest point of attachment 200 mm.

A dark-green glass dome, the bottom edge of which is turned inwards and upwards to form a gutter to which a glass spout has been attached, sloping away at an angle. A thick knob of glass attached to the summit of the dome acts as a handle. (see figure 48)

57-60 Four flasks
[Globular bottles (4 specimens). 1858.275.43]

(57) height 340, maximum diameter 260; (58) height 290, maximum diameter 210; (59) height 270, maximum diameter 170; (60) height 230, maximum diameter 170 mm.

Flasks (57) and (59) are of pale-green, (60) is of dark green and (58) is of thick dark-green glass. (57), (58) and (60) are flat-bottomed and have unground pontil-marks. (59) is round-bottomed. (58) has a prominent, rolled-over rim.

Figure 53 Flasks (object 57, left; object 58, second from left; object 59, third from left; object 60, right)

61. Bell-shaped vessel
[Large bell shaped vessel. 1858.275.44]

Height 290, base diameter 305 mm

Pale-green glass vessel of a bell-shape, the glass containing large air bubbles, brought to a spout (broken off) at the top. The bottom edge has been carefully folded over to form a rim.

Figure 54 Bell-shaped vessel

62, 63 Two tubes for producing electrostatic charges
[Two Electrophorus handles. 1858.275.45]

(62) length 410, approximate diameter 45; (63) length (of both parts placed together) 505, approximate diameter 45mm.

Tubes of thick dark-green glass, sealed at one end and broken off at the other. (63) has been broken and is in two parts, break approximately 120 mm from the closed end.

Figure 55 Tubes for producing electrostatic charges. (object 62, below; object 63, above)

The collection of 24 pieces of green glass in the Playfair Collection has been known as 'Black's Glass' since it was acquired by George Wilson, Director of the Industrial Museum of Scotland, in 1858. Amongst his acquisitions for that year, Wilson remarked in his annual report: "In glass, several examples of old Scotch glass have been obtained from the laboratory relics of Dr Black".[1] In the catalogue entry of the Museum Register, Wilson indicated a possible source:

> "various forms of glass vessels used by Dr Joseph Black between 1766 and 1799. They are supposed to have been made at Leith."[2]

No definite evidence survives (if any ever existed) to associate the chemical glassware with Black, or to suggest its original source. Glass of this type was available from the time from when chemistry teaching started in Edinburgh, at the beginning of the 18th century, to the date of acquisition of the collection by Wilson. There is no positive reason to associate it with the glassworks at Leith, except that they were geographically the nearest. Furthermore there is no reason why all items of the collection should date from the same period. Unlike certain items of chemical apparatus which, in common with philosophical and mathematical instruments, can be identified and dated by reference to published papers, catalogues, other similar pieces of good provenance, etc. glassware does not lend itself to this treatment except under exceptional circumstances. The technology of glass manufacture slowly evolved, but changes affected the more expensive types of product. There is nothing characteristic about the design or form of the crude Playfair Collection glass. It has been suggested that pontil marks (lumps of rough glass left after the vessel was snapped off from the iron rod, or pontil, which held it during its manufacture) which have been ground down may indicate a post-1860 date[3] but this does not apply in dating this glass.

Because glass has a non-crystalline structure, no analytical technique analogous to the thermoluminescent dating of ceramics can be invoked. It is of interest, however, that Wilson supervised the chemical analysis of a piece of Black's Glass soon after it was acquired. The investigation was conducted by Thomas Bloxam, Assistant Chemist to the Industrial Museum of Scotland. The paper (which included the results of analyses of five other pieces of glass of interesting derivation) was read to the Royal Society of Edinburgh in March 1859. The preamble by Wilson includes the following remarks:

> "Of dark bottle-glass, specimens abundantly authentic were obtained from the relics of Dr Joseph Black's apparatus. They were probably made at Leith, where ordinary bottles have been manufactured for a long period from clay, sand, salt and other cheap materials in the neighbourhood. Newcastle, however, has long directed much of the bottle making to herself, and still more recently, Belgium is injuring the trade at Newcastle."[4]

The results of Bloxam's analysis reveal nothing of great interest and indeed the exact reason for performing the investigation is not clear.

The earliest surviving chemical glassware of British origin dates from the mid-15th century.[5] The first steady demand for glass chemical apparatus in Scotland probably came from John Damion ('French Master John') who was installed in King James IV's alchemical laboratory at Stirling from 1501. From 1502 to 1508 there are ten entries in the accounts of the Lord High Treasurer of Scotland for glassware clearly intended for his chemical or pharmaceutical use.[6] The usual entry was for *flacatis* (large flasks) or *urinales* (small flasks). It is likely that glass used in Scotland in the 16th century for chemical purposes would have been imported from the Continent, as it was in England[7], though glassmakers were active at Tyneside, not too far away, from the middle of the century.[8] Scottish made glass was available from 1610 when Sir George Hay of Nethercliff was granted a licence to make glass for a period of 31 years.[9] His works were set up at Wemyss Bay, Fife. From the mid 17th century glass was manufactured in Scotland almost without a break, the main centres being Leith, Glasgow and Alloa.[10] By 1689 it was stated that Leith was able "to undertake clear chemistry and apothecary wares in greater quantity in four months than was vended in the Kingdom previously in one year and at as low a rate of goods of a similar quality from Newcastle or London". Daniel Defoe recorded in 1727 that Newcastle was supplying chemical ware of flint glass[12] and three years later the York Buildings Company glasshouse at Port Seton (to the east of Edinburgh on the Firth of Forth) advertised that "Glasses for Alchymists" were available.[13]

The first steady demand for glassware for a university laboratory in Scotland came from John Rutherford, Andrew St Clair, Andrew Plummer and John Innes of Edinburgh. In February 1726 they were appointed professors of medicine of the new medical school (though they had been teaching from 1725) and in September 1726 they started ordering large quantities of chemical glassware and drugs from London. The scale of their operation indicates that they were producing pharmaceutical products as large scale suppliers to apothecaries. The glassware was ordered mainly from Glisson Maydwell of The Strand (orders from September 1726 to January 1741) and from Alexander Johnston (from August 1733 to December 1735).[14] The partnership was dissolved in March 1742. It may be assumed that glass was not available locally in the quantity or quality required. This is also a possible conclusion drawn from a letter sent to William Cullen (lecturer in chemistry at Glasgow) from his cousin, Walter Johnson, in London.[15] Cullen had had to order tubulated retorts and flasks from London and Johnson was apologising for the delay, saying "I have found great Difficulty to get any Body to undertake blowing them as They are seldom or never us'd here by any of the Chymists". The earliest surviving list of chemical glassware at Glasgow, an inventory titled 'List of the Utensils delivered to the Committee appointed to receive the Laboratory from Dr Black', dated 25 October 1766, separates 'white Glass' from 'Green Glass'.[16] A later inventory includes a note made by William Irvine in May 1773 when Joseph Black reclaimed some glass apparatus which he had left in Glasgow seven years earlier. Irvine records "Upon the thirteenth day of July last Dr Black took away two lapdals for Cucurbits & two Separatories contained in the above list; he said that they were part of some Glass Utensils left by him in the Laboratory".[17] Again this evidence suggests that chemical glassware was not readily available at this time in Scotland.

It is possible that it was Black who helped to remedy this situation. One of his brothers worked for (and in 1798 was Secretary of) the British Cast Plate Manufactory in London (though the factory itself was at Ravenhead, St Helen's). A series of letters between Joseph and Alexander discusses technical matters concerning the production of glass.[18] Perhaps Black's closest association with an industrialist was with Archibald Geddes, manager of the Edinburgh Glass-house Company at Salamander Street, Leith.[19] Black's involvement with Geddes and the Company will be considered in some detail because it was felt in the mid-19th century that this was the most probable source of the Playfair Collection glass.

Initially the relationship between Black and Geddes was a formal one, Geddes attending Black's lectures, though it later developed into one of close friendship. As John Robison recorded in the Preface to his edition of Black's lectures:

> "Soon after his coming to Edinburgh, the Doctor [Black] got another pupil, Mr Archibald Geddes, manager of the glass-works at Leith, who soon engaged his Professor's attention by the readiness and propriety with which he applied to the improvement of his manufacture the instructions which he received in the lecture. Farther acquaintance shewed more to esteem and attach; and it terminated in the most intimate and confidential friendship. From this friend no circumstances of Dr Black's former life or present condition was withheld; and to his assistance he had recourse in every thing that affected either his fortune or his comfort."[20]

It seems quite likely that Black was influential in the Edinburgh Glass-house Company diversifying its products and producing flint glass. Until at least 1782, Geddes' Company was producing only glass of bottle quality (the lowest grade, made from sand and lime with small quantities of soda, iron impurities colouring it green).[21] But around this date the glasshouse at Leith started to make better quality glass, possibly for pharmaceutical use.[22] By 1787 the Leith works was definitely producing chemical glassware of some complexity. The evidence appears in a pamphlet advertising a Nooth's apparatus for the preparation of aerated water:

> "GLASS MACHINES, of different sizes and constructions, for preparing Medicinal Waters, and all sorts of Glasses for Chemical and Philosophical Experiments, made by The Edinburgh Glass-house Company, at their Bottle, Flint, and Window-Glass Manufactory, Leith.
>
> ARCH GEDDES Manager"[23]

An advertisement for pharmaceutical glassware also appears in the 1792 edition of the Edinburgh Pharmacopoeia[24] issued by the Royal College of Physicians of Edinburgh (Black played a major role in its revision).[25] By 1797 the Edinburgh Glass-house Company was offering a wide range of scientific glassware in flint and green glass in its price list.[26] There is one further, and important link between Joseph Black and the glassworks: on 1 December 1794 Black purchased five bonds of £100 each in favour of the Edinburgh Glass-house Company from Archibald and William Geddes.[27] Dividends on these were paid on 14 and 27 December 1796 and 31 December 1798.[28]

It seems probable that Black would obtain his glassware from the glassworks managed by his friend Geddes, and it is possible that he promoted the production of chemical ware because of the difficulty in obtaining supplies in Scotland. As late as 1784 Black had to advise that such glass be obtained from London.[29] Thomas Charles Hope continued to patronise the Leith works after Black's death. In 1805 Benjamin Silliman, the American chemist and geologist, visited Hope in Edinburgh and the men visited the glassworks at Leith, a visit recorded by Silliman:

"To Dr Hope I was indebted for other civilities, particularly in walking with me to Leith, to use his personal influence in obtaining for me some articles of glass apparatus, especially some instruments like those which I had seen successfully used in his own experiments."[30]

One other local glassworks which might have supplied Hope was the Mid-Lothian Flint Glass Works at Portobello, close to Leith on the Firth of Forth. There is a record of Hope's assistant David Boswell Reid taking parties of students to see glassware being made there. "The workmen having been directed to stop the operations they were then engaged in, and make all the more important varieties of chemical apparatus . . . added much to the interest of these visits."[31] Hope certainly obtained glass from works in the Canongate, Edinburgh, and from Glasgow (see chapter 3, ref 69).

The Playfair Collection glass would seem to have originated in a bottle works and does not appear to be of the quality which would have been available from specialist dealers in the 19th century. Most is thick and hence liable to crack when heated. The retorts are inexpertly made with constricted necks. Black taught at his lectures that green glass was more likely to crack or break than flint glass.[32] Possibly some pieces from the Playfair Collection glass date from when Black was forced to use apparatus made by workmen used to making bottles, there being no men experienced in blowing chemical glass available. However Black seemed happy to utilise what domestic vessels were at hand indicating, perhaps, discrimination against more sophisticated pieces.[33] Less, if anything, is known about Hope's glass apparatus, except that some of it was large. This was specially remarked on by a visitor to his laboratory in May 1818: "his apparatus, in general, is upon a large scale; in the laboratory he has reservoirs of the different gases . . . he has a glass reservoir, capable of holding about ten or twelve gallons of oxygen".[34] Gregory's research which was of tradition of German organic chemistry (he had studied with Justus Liebig at Giessen in 1835-6). It is most unlikely that the crude glassware in the Playfair Collection was acquired by him. The question as to who did actually acquire it, and from what source, is one which will probably not be resolved. The pieces of thick dark green glass containing air bubbles are likely to have been made in a glassworks conditioned only to making bottles.[35] The better made vessels of thinner glass may possibly date from a later period and may have been supplied by a glasshouse which specialised to some extent in these products.[36]

The glassware in the Playfair Collection is not altogether typical of those very few groups of contemporary chemical glass which survive and which are of certain provenance. Two such groups, one in the Royal Institution of London and the other in the Museum of the History of Science, Oxford, both of the first half of the 19th century, include vessels of clear glass which are 'quilled' (ie which have tapering spouts to join to necks of other vessels) and which have blown-on tubulatures.[37] None of the Playfair Collection items has these refinements. However several pieces are closely similar to apparatus illustrated by Robison in his published edition of Black's lecture notes.[38] The Playfair Collection glass cannot be compared with the complex demonstration apparatus made for Antoine Lavoisier and Martinus van Marum in the late 18th century which was produced for a different function.

The very large ellipsoidal flasks (objects 45, 46) may possibly be associated with Hope considering his predilection for large glass apparatus[39] and their finer construction perhaps implying a later date of manufacture. Of the other flasks, the long-necked variety (object 50, usually called a Florentine flask) was particularly referred to by Black, both in his lectures[40] and in his research.[41] The alembic and retorts (objects 52-54, 56) were standard apparatus for distillation operations in the 18th and 19th centuries. In both forms the distillate condensed on the cool upper concave surface; in a retort the distilled liquid flowed straight out of the spout, in an alembic the liquid first ran into an internal gutter before being led out through the spout.

The bell-shaped vessel (object 61) was probably used in the preparation of gases, being placed on a flat surface over reacting chemicals and having a tube luted to its spout which would pass into a pneumatic trough.[42] Though the lack of a stoppered inlet would be a severe drawback to its practical use, the suggestion that it was used as a funnel is less likely on account of its shape, size and carefully folded over rim.[43]

The two glass tubes were probably made for the generation of static electricity. The Library Company, Philadelphia, was presented with a glass tube for making electrical experiments and directions for using it by Peter Collinson (1694-1768) in 1764.[44] It was Collinson who stimulated Benjamin Franklin with his "new German Experiments in Electricity". Franklin "caused a number of similar tubes to be blown at our glasshouse" and describing these, said "our tubes are made here of green glass, 27" or 30" long, as big as can be grasped".[45] A similar tube survives in the collections of the American Philosophical Society at Philadelphia.[46] Black may have discussed these experiments when he and Franklin dined together in Edinburgh on 18 November 1771.[47] It is curious, considering the inclusion of these tubes in the Playfair Collection, that Black wrote: "the instruments for exciting electricity would be improperly considered as part of the chemical apparatus."[48]

It has been mentioned elsewhere that the glass possesses an aesthetic appeal.[49] Perhaps this quality was appreciated by Black, with reference to whom Robison commented "Even a retort, or a crucible, was to his eye an example of beauty or deformity . . . These are not indifferent things; they are features of an elegant mind."[50]

NOTES AND REFERENCES

1. *Sixth Report of the Science and Art Department of the Committee of Council on Education* (London 1859) Appendix F. The glass has previously been listed and briefly described, see Revel Oddy 'Some Chemical Apparatus blown by hand in the late 18th to early 19th century' *Annales du 5e Congrès de l'Association Internationale pour l'Histoire du Verre* (Liege 1972) 225.

2. Royal Scottish Museum, Edinburgh, MS Museum Register (1855-1858), entry for 30 September 1858. Leith is Edinburgh's seaport two miles distant.

3. J K Crellin and J R Scott *Glass and British Pharmacy* (London 1972) 19.

4. Thomas Bloxham 'On the Composition of Old Scotch Glass' *Proc Royal Soc Edin 4* 191 (1862).

5. Stephen Moorhouse 'Medieval Distilling-Apparatus of Glass and Pottery' *Medieval Archaelogy 16* 79 (1972).

6. J B Paul (ed) *Compota Thesaurariorium Reguum Scoutorum 2-4* (Edinburgh 1900-1902). For entries in the accounts for chemical glassware, see *2* 20 May 1502, 30 October 1503, 23 November 1503, 18 January 1504; *3* 13 September 1505, 9 May 1506, 13 August 1506; *4* 29 November 1507, 24 December 1507, 12 May 1508. James IV's alchemical activities are discussed in *ibid 2* lxxiii - lxxix and in John Read *Humour and Humanism in Chemistry* (London 1947) 18.

7. E S Godfrey *The Development of English Glassmaking 1560-1640* (Oxford 1975) 247.

8. Ursula Ridley 'The History of Glass Making on Tyneside' *The Circle of Glass Collectors Paper 122* (1961).

9. Godfrey *op cit* (7) 97. The licence was ratified in 1612; see T Thomson (ed) *The Acts of the Parliaments of Scotland (1124-1707) 4* (Edinburgh 1816) 515.

10. The development of the glass-making industry in Scotland has not been adequately traced. However the following is a list of glasshouses operating before 1800 recorded by J Arnold Fleming *Scottish and Jacobite Glass* (Glasgow 1938), Revel Oddy (1) 'Scottish Glass Houses' *The Circle of Glass Collectors Paper 151* (1966) and Revel Oddy (2) *op cit.* (1).

Place	Founder/Manager	Earliest Date	Reference
Wemyss Bay	George Hay	1610	Fleming 95, Oddy (1) 2
Prestonpans	William Morrison	pre-1661	Fleming 101, Oddy (1) 4
Leith, Citadel	Robert Pape	pre-1664	Fleming 110, Oddy (1) 4
Leith	Charles Hay	1682	Fleming 111
Leith	Robert Douglas	1695	Fleming 111
Leith	William Scott and Paul le Blanc	ca 1699	Fleming 111
Glasgow, Jamaica Bridge	James Montgomerie	ca 1701	Fleming 101
Port Seton	York Buildings Company	ca 1720	Oddy (1) 4
Alloa	Frances Erskine	1750	Oddy (1) 4
Glasgow, Verreville	Patrick Colquhoun	1770	Fleming 130
Middle Leith	James Ranken	1773	Oddy (1) 5
Dumbarton	Jacob Dixon	1776	Fleming 155
Dundee		1789	Oddy (2) 227
Greenock	John Geddes	1793	Oddy (2) 227

11. Fleming *ibid* 111.

12. Daniel Defoe *The Complete English Tradesman 2* (London 1727) 63.

13. *Edinburgh Evening Courant* 9 February 1730 ("That at the Glasshouse at Portsetton there is to be sold . . . all sorts of Flint or Chrystal Glass . . . Vials &c Glasses for Alchymists"). For details of the Port Seton Glassworks see David Murray *The York Buildings Company. A Chapter in Scotch History* (Glasgow 1883) 66, 73.

14. 'Book of Letters and Envoys Belonging to the Elaboratory', manuscript volume in private collection (Miss A Scott-Plummer). The first order to Glisson (15 September 1726) was for 105 pieces of glass (including six 3 gallon retorts and six 4 gallon receivers) at a cost of £10 13s 6d.

15. Glasgow University Library MS 2255/27, letter Johnson to Cullen, 8 December 1747.

16. Glasgow University Archives MS GUA 43081. The inventory is signed by John Robison as having received the contents of the laboratory.
17. Glasgow University Archives MS 'List of the Utensils in the Laboratory as delivered to Dr Irvine July 13 1769'.
18. Douglas McKie and David Kennedy 'On some Letters of Joseph Black and others' *Annals of Science 16* 129 (1962). Ten of the 26 letters published from Joseph to Alexander Black deal directly with glass manufacture.
19. Archibald Geddes is not easily traced. He appears to have been one of a family several of whose members were associated with glass (see Fleming *op cit* (10) 112, 113). He was admitted a founder-member of the Chamber of Commerce and Manufactures of Edinburgh on 10 January 1786 (see *Minutes and Proceedings of the Chamber of Commerce and Manufactures at Edinburgh* (Edinburgh 1788)). Geddes was manager of the glassworks during a period of rapid expansion of the industry at Leith, despite heavy (and intricate) excise regulations (see R W Douglas and S Frank *A History of Glassmaking* (Henley-on-Thames 1972) 30) suggesting that he was an able industrialist.
20. Joseph Black *Lectures on the Elements of Chemistry . . . Published by John Robison 1* (Edinburgh 1803) vi. See also Black's obituary written by Adam Ferguson *Trans Royal Soc Edin 5* 101 (1801) ("Mr Geddes of the Glass Works at Leith, an eminent manufacturer who knew well the value of his friend's suggestion"). In a letter to his brother George (dated 21 August 1793), Joseph Black wrote Geddes an introduction: "This will be delivered to you by Mr Geddes the manager of the principal Glassworks at Leith who has long been my most intimate acquaintance & one of my best friends" (Edinburgh University Library MS Gen 874 *vol 5* f62).
21. McKie and Kennedy *op cit* (18) 136. Despite fiscal discouragement (see Douglas and Frank *op cit* (19) 30) the glass industry at Leith appears to have flourished in the late 18th century. From July 1777 to July 1778, 15,883¼ cwts (approx 80700 kg) of glass was manufactured at Leith, the duty payable on which was £2779 11s 4¼d (see Hugo Arnot *The History of Edinburgh* (Edinburgh 1788) 587).
22. *The Bee* (29 August 1792) 333 ("the glasshouse company at Leith confined their efforts till about a dozen years ago, when they began to make fine glass for phials and other articles on that nature").
23. *Directions for preparing Aerated Medicinal Waters* (Edinburgh 1787). This apparatus, first described by John Mervin Nooth (see *Phil Trans Royal Soc 65* 59 (1775)) was in vogue at the end of the 18th century. A similar pamphlet published seven years later, John Moncrieff *An Enquiry into the Medicinal Qualities and Effects of the Aerated Alkaline Water* (Edinburgh 1794), advertised "Glass machines for preparing Aerated Alkaline Water" as being available from the same company.
24. D L Cowen 'The Edinburgh Pharmacopoeia' in R G W Anderson and A D C Simpson (eds) *The Early Years of the Edinburgh Medical School* (Edinburgh 1976) 34.
25. *Pharmacopoeia Collegii Regii Medicorum Edinburgensis* (Edinburgh 1792) ("Glass Measures for Wine, Water and Watery Fluids, accommodated to the Troy Weight; and Phials of a particular shape . . . are made and sold by the Edinburgh Glass-House Company"). There is evidence that Black influenced the design of pharmaceutical phials: the statement "The late Dr Black, with a view to obviate the inconvenience arising from this vague and indefinite manner of ascertaining small doses of tinct of opium, &c proposed that the lips of the phials kept in the apothecaries shop for this purpose should be ground to a certain thickness." appears in 'Dr Charles Anderson's Machine for measuring small quantities of Fluids' *Edinburgh Phil J 8* 418 (1823).
26. *Prices of Flint Glass Manufactured by Edinburgh Glass-house Company* (Leith 1797). This list cannot now be traced, though a photograph of it is held by the Royal Scottish Museum. Glassware advertised includes the following: Electrical Cylinders (ie glass cylinders for electrostatic machines), Chymical Glasses, Phials (white or green glass), Retorts and Receivers (green or flint glass), Barometer and Thermometer Tubes, Urinals (green or flint glass).
27. Edinburgh University Library MS Gen 874/XI ('Copy of the Minutes taken at the opening of the Repositories of the late Dr Joseph Black'); "There was found in the Cabinet of the Dr's Bed Room . . . 5 Bond by Messrs Archd & Wm Geddes of the Leith Glass House for £500 St in favour of the said Dr Black dated 1 Decr 1794".
28. Edinburgh University Library MS Gen 874/X ('Account begun 31 Augst 1792 of Interest of money dividends & some other sums received by me'). The income received on 31 December 1798 is entered as follows: "Received from Mr Archd Geddes... £25 with deduction of the price of a cow £7 which he bought for me in November last . . . £18". (Could it be that the famous glass of milk balanced by Black on his knee when he died on 6 December 1799 came from this very cow purchased from the income derived from the Leith glassworks?).
29. Edinburgh University Library MS Gen 873 volume 2, 162F ("Mr Hutchins must procure the Glasses from London as there are none to be had in Edinburgh").
30. George P Fisher *Life of Benjamin Silliman M.D., LL.D.1* (London 1866) 161.
31. *Testimonials regarding Dr D.B. Reid's Qualifications as a Lecturer on Chemistry* (Edinburgh 1833) 38. This reference occurs in a published testimonial in favour of Reid when he was hoping to be appointed to a newly established post of practical chemistry. This particular passage occurs as a note by Reid accompanying the testimonial of the proprietors of the works, Messrs Bailey and Company.
32. Many sets of lecture notes taken by Black's students make this point. See, for example, Glasgow University Library MSS Gen 49/1, f258 and Ferguson 41, f47.
33. In 1771 Black recorded using "a cylindrical glass-vessel, about six inches in diameter, such as confectioners have in their shops" (see Francis Home *Experiments on Bleaching* 2nd ed (Dublin 1771) 269 (Appendix by Joseph Black 'An Explanation of the Effect of Lime upon Alkaline Salts')); in 1775, he records "putting four ounces of boiled and four of the unboiled water, separately, into two equal tea cups . . ." (*Phil Trans Royal Soc 65* 124 (1775)); in 1785 for an analysis of impurities in a water supply Black used "beer Glasses" from which he withdrew "from each water a Wine glassfull" (Edinburgh University Library MS Gen 874 ff 20*, 20**). A MS illustration of apparatus drawn by Black for a linen bleaching process (c 1763)

(Edinburgh University Library MS Gen 873 f8) shows a very simple arrangement, as does a figure of Black's apparatus published by David Macbride *Experimental Essays* (London 1764) 51. On the other hand, the glass apparatus painted by David Martin in his portrait of Black (of 1787) shows well-made white glass of a moderately sophisticated nature (portrait commissioned by, and still the property of, The Royal Medical Society, Edinburgh).

34. *Life of William Allen with Selections from his Correspondence* (London 1846) 355.

35. Playfair Collection, objects 40-44, 48-54, 56-58, 60-63.

36. Playfair Collection, objects 45-47, 55, 59.

37. Glass apparatus associated with C G B Daubeny, now at the Museum of the History of Science, Oxford, is illustrated in R T Gunther *Early Science in Oxford 1* Part I (Oxford 1920) 83.

38. Black *op cit,* (20) *2* plate II.

39. Allen *op cit* (34).

40. Black *op cit* (20) *1* 288 ("Florence flasks are extremely serviceable in such operations").

41. Joseph Black 'Experiments upon Magnesia Alba, Quick-lime and other Alcaline Substances' *Essays and Observations, Physical and Literary 2* 157 (1756) ("I chose a Florentine flask, on account of its lightness, capacity, and shape").

42. A vessel similar in shape is described and illustrated being used in this manner in [Friedrich Accum] *An Explanatory Dictionary of Chemistry* (London 1824) 247, plate II figure 18.

43. Oddy *op cit* (1) 230.

44. I Bernard Cohen *Benjamin Franklin's Experiments* (Cambridge Mass 1941) 18; William B Willcox (ed) *The Papers of Benjamin Franklin 17* (New Haven 1973) 66 (letter from Franklin to Collinson, 8 February 1770). The tube still survives at the Library Company: it is of green glass, 33 inches long and 1¾ inches wide, its surface roughened by having been ground with an abrasive powder.

45. Benjamin Franklin *Experiments and Observation on Electricity* (London 1769) 11.

46. Robert B Multhauf *A Catalogue of Instruments and Models in the Possession of the American Philosophical Society* (Philadelphia 1961) 19, figure 9.

47. Ernest Campbell Mossner *The Life of David Hume* (Oxford 1970) 573.

48. Black *op cit* (20) *1* 303. This passage may be due to Robison and not Black.

49. Oddy *op cit* (1).

50. Black *op cit* (20) lxvii.

64. Instantaneous light machine
[Arrangement for producing a light by igniting a jet of hydrogen gas by means of a spark from an electrophorus. 1858.275.46]

Overall height 322, base 228 x 187, diameter of cylinder 162 mm.

A tinned iron cylinder, bronzed on the exterior and painted on the interior, is mounted by means of a metal collar on to a rectangular wooden (beech) base, supported at each corner by spherical brass feet. The box has a removable panel in the front which reveals an electrophorus (of dimensions nearly those of the box itself) constructed from two plates, one fixed, the upper hinged, coated with red wax. A small cavity at the back of the box stores a rectangular piece of zinc. The cylinder is divided horizontally into a lower and upper chamber, the latter having fixed at its circumference a small brass cylinder capped with a brass plug with a rectangular grip. Attached to the top of this small cylinder a U-shaped tube communicates to the lower cavity. An ivory key attached to the side of the box operates the upper electrophorus plate and a valve. In operation, water contained in the upper chamber forces hydrogen produced in the inner cylinder (by the action of sulphuric acid on zinc) through a small concealed jet. It is ignited by means of a spark from the electrophorus and lights an oil lamp (in the shape of a cruse) fixed to the front panel of the base.

Figure 56 Instantaneous light machine

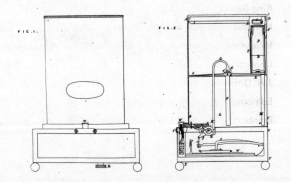

Figure 57 Instantaneous light machine of Matthew James Meyer from patent specification of 1811

This apparatus produces a flame by igniting a stream of hydrogen. The first arrangement of this *genre* was described in 1780 by Friedrich Ludwig Ehrmann (1741-1800) whose apparatus produced a hydrogen jet which was lit by means of a spark.[1] In 1807, Richard Lorenz took out a British patent for 'Producing Light and Fire Instantaneously'.[2] This contrivance generated hydrogen by means of the action of sulphuric acid on zinc. A head of water forced the hydrogen through a nozzle where it was ignited by means of an electric spark from an electrophorus (previously charged) situated in the base of the instrument. By enclosing the mechanism in a wooden box, the effect was intended to be spectacular. In 1811 Matthew James Mayer made minor changes to Lorenz's apparatus and this modification (fig 57) was also granted a patent.[3] The main advantage seems to have been that Mayer's instrument could be used when subjected to "any common motion to which the machine may be exposed on board of ships, or in carriages". In practice it seems that few were made, and it is unlikely that they were used under such utilitarian conditions.[4]

The Playfair Collection instrument is a Mayer-type Instantaneous Light Machine. A very similar example in the University of Glasgow has an attached plate with 'MAYER'S PATENT' embossed on it.[5] These instruments are unlikely to have been manufactured after 1823 when the much more convenient lamp produced by J.W. Döbereiner was described.[6] This ignited the hydrogen when it was passed over a catalyst of spongy platinum.

Hope demonstrated both Mayer's instrument (which he called Volta's electric lamp) and Döbereiner's lamp to his students:

> "*Two very* different, but equally beautiful contrivances for affording Instantaneous Light have been contrived.
> *In the one*, a small stream of Hydrog. g. is set on fire by an Electric Spark;
> *In the* other by the contact of Platinum —
> *This* is Volta's Electric Lamp — When I turn this Key — The Lamp is kindled I turn the Key, I open the Stopcock, & the pressure of Water in the Upper Cavity forces out a quantity of gas, & at the same instant an Electric Spark proceeds from the Electrophorus, darts thro' the Stream & kindles it —"[7]

As might be expected, Hope did not miss the opportunity of demonstrating this intriguing apparatus to his classes of popular lectures in 1826 and 1828.[8]

NOTES AND REFERENCES

1. F L Ehrmann 'Description et Usage de quelques Lampes à Air Inflammable' (Strasburg 1780). For a discussion of the development of some instantaneous light contrivances, see Knud E Holm 'Nolge Fyrtøjer fra 17— og 1800-tallet og deres Forudsaetninger med en Excurs til Kipp's Apparat' *Nationalmuseets Arbejdsmark* (Copenhagen 1971) 113. There are examples of Ehrmann-type instruments in the Danish Technical Museum (Holm *ibid* fig 14) and in the Science Museum, London (inv. no. 1950-232). There is an extensive collection of instantaneous light machines at the Deutsches Museum, Munich.
2. Richard Lorentz 'Producing Light and Fire Instantaneously' British Patent 3007 (1807). A Lorentz-type instrument survives at the Science Museum, London, in the King George III Collection (inv. no. 1927-1181), see J A Chaldecott *Handbook of the King George III Collection* (London 1951) 75, item 306.
3. Matthew James Mayer 'Instantaneous Light Machines' British Patent 3470 (1811).
4. Miller Christy *The Bryant and May Museum of Fire-Making Appliances* (London 1926) 93.
5. University of Glasgow, Natural Philosophy Department Collection (inv. no. M.9).
6. J W Döbereiner 'Neu entdecke merkwürdige Eigenschaften des Platinsuboxyds' *Ann Phys* **74** 269 (1823).
7. Edinburgh University Library MS Gen 271, envelope 150 ("Hydrogene Water 3d").
8. Edinburgh University Library MS Gen 271, envelope 126 ("Popular No 3d").

APPENDIX 1

The following items which appear in the Museum Register entry of 30 September 1858 cannot now be located, or have been destroyed.

i. **Differential Thermometers or Thermoscopes (3 specimens). 1858.275.10**
   [One specimen missing]

See notes for the two surviving differential thermoscopes (objects 10 and 11).

ii. **Double Registering Thermometer (early form) by Adie, Edinr. 1858.275.10a.**
   [Destroyed, Minute of Disposal Board, 4 April 1946]

Thermometers which recorded the maximum and minimum temperature over a period of time (i.e. double register thermometers) were proposed from the end of the 17th century, but the first to incorporate indices moved by the column of mercury was invented by a Scot, John Rutherford, a century later. Details of it were transmitted to the Royal Society of Edinburgh by Daniel Rutherford, professor of medicine and botany, in April 1790.[1] The maximum thermometer (which originally had a conical ivory index, later replaced by one of steel) was somewhat unsatisfactory as the mercury tended to flow past the index. The Edinburgh scientific instrument maker Alexander Adie (who set up in business with his uncle in 1804[2]) remedied this fault by introducing naphtha above the index; the date of this improvement was suggested in 1859 to have been made between about 1829 and 1839.[3]

Adie himself kept regular meteorological records, initially at his home at Canaan Cottage and later at Calton Hill, Edinburgh, using "very nice instruments, constructed by himself". The readings included maximum and minimum temperatures and his records were published in the *Edinburgh Journal of Science* between 1824 and 1832. However, remarking on an article on maximum thermometers by James King of New South Wales, Adie (writing under the cryptonym '△'), commented "The [register thermometer] which I daily use is by one of the first London makers, yet it is frequently out of order, and follows imperfectly the atmospheric changes. Much, very much, is to be looked for in the improvement of this thermometer".[4] Adie was clearly interested in improving the instrument by the date of this remark (1828) but had not yet done so. However a manuscript written by William Gregory, 'Outlines of Lectures on Chemistry Begun October 1828' contains the comment 'Register thermometers. Those of Dr Rutherford improved by Adie are the best.'[5] Thus it may be assumed that Adie's version dates from about this time.

Rutherford's thermometer and its variants were, in fact, two independent thermometers mounted side by side. A different type of maximum and minimum thermometer in which both functions were combined in a single instrument was invented by James Six in 1782.[6] This did not prove very satisfactory and its shortcomings were discussed in an article by Alexander Adie's son, Richard (also an instrument maker), in 1853.[7] While it cannot be certain that the Playfair Collection thermometer was of the Rutherford type, inclusion of the comment '(early form)' probably implies that it was of the variety that Alexander Adie was interested in improving rather than that associated with his son. It is known that Thomas Charles Hope demonstrated a Rutherford type thermometer to his class.[8]

1. Daniel Rutherford 'A Description of an Improved Thermometer' *Trans Royal Soc Edinburgh 3* 247 (1794).
2. D J Bryden *Scottish Scientific Instrument Makers 1600-1900* (Edinburgh 1972) 9, 14, 43, 54. It is perhaps surprising that the Playfair Collection did not include more apparatus by Adie, who was a highly competent and original instrument maker.
3. John G Macvicar 'Notice of another New Maximum and Minimum Mercurial Thermometer' *Quarterly J Chem Soc 11* 106 (1859).
4. 'Remarks on Self-Registering Thermometers, particularly on Mr King's' *Edinburgh J Sci 9* 300 (1828). The thermometer which Adie was using was of Six's type.
5. Aberdeen University Library MS 2206/25.
6. James Six 'Account of an Improved Thermometer' *Phil Trans Royal Soc. 72* 72 (1782).
7. Richard Adie 'On an Improvement in Sikes' Self-Registering Thermometer' *Edinburgh New Phil J 54* 84 (1853). Richard Adie is recorded as working as a scientific instrument maker in Liverpool from 1835 to 1875.
8. Edinburgh University Library MS Gen 270, List of Specimens (notebook compiled by Hope for his assistant's use in preparing demonstration apparatus). Included is the entry: 'Register Thermr 1t, Rutherfords & Diag. 2d Barbons & Drawing - 3r Keiths & Drawing - Crichtons - Pocket Do'.

iii. **Stock of a pistol used by Professor Hope for shewing the mode of firing percussion caps when they were first introduced. 1858.275.14.**

The percussion lock (in which a hammer hits a striker which detonates a small charge of explosive[1]) superseded the flint-lock and was introduced and developed by the Reverend Alexander John Forsyth of Belhelvie, Aberdeen, between 1805 and 1807.[2] The percussion cap in which the explosive (originally mercury fulminate) was contained in a (usually) copper cap was first produced ca 1816-1820. There were many claimants to its discovery.

It is likely that the entry in the Museum Register is in error and should have been 'percussion lock' rather than 'percussion cap'. It is certain that Thomas Charles Hope demonstrated Forsyth's lock prior to 1812. His notes from which he lectured read:

> "The Revd Mr Forsyth uses a Detonating powder made with it *as priming* in his beautiful Invention of the patent Gun Lock, in which it is kindled by Percussion -
> *This* is the contrivance
> *The Pan* for the priming & communicates with the touch hole is very small - & not more than ½ grain of the powder is used. This powder is composed of
> 6p S.ox.m. 1-Charcoal 1-Sulphur
> It is inflamed with percussion
> You see a steel pin, which when struck by the Dog head enters into the little pan & kindles the powder -
> *Expt with Lock*".[3]

1. J F Hayward *The Art of the Gunmaker 2* (London 1963) 320.
2. Alexander John Forsyth Reid *The Reverend Alexander John Forsyth . . . and his Invention of the Percussion Lock* (Aberdeen 1909).
3. Edinburgh University Library MS Gen 269 envelope 65 ('Old Notes Oxd Mur Pota Not Used'). Included in the bundle of papers is a sheet which begins "Mr Innes's improved lock" and the pencilled date 1812, the first year in which Hope added these additional notes to his lectures on Forsyth's percussion lock.

iv. Experimental Oil Lamp. 1858.275.18

Though no details are given by the register entry, it is likely that this object was a variety of Argand lamp, many types of which were developed in the last decade of the 18th century and in the early part of the 19th century (see object 17).

v. Sectional Model of a Reverberatory Furnace. 1858.275.20

In the reverberatory furnace the products of combustion are deflected downwards by a sloping roof towards the hearth on to the metal ore which is being reduced. The furnace is described in the 16th century metallurgical treatises.[1] In 1784 Henry Cort patented a method of converting cast iron to wrought iron by means of a reverberatory furnace.[2] Joseph Black took a close interest in Cort's work and it seems that he commissioned small laboratory reverberatory furnaces from the Carron Iron Works in Stirlingshire[3], possibly of a French pattern described in 1776.[4] Among the figures used by John Robison in his published edition of Black's lecture notes is a representation of the experimental furnace devised by Peter Macquer. This was not mentioned in the text though for some reason Robison was at pains to include it.[5] A "Model of a Furnace" was shown by Thomas Charles Hope to his students and this may be the example which was once in the Playfair Collection.[6] It is impossible to speculate the exact form this item took. It seems likely that Black, with his interest in furnace construction, may have first demonstrated it to his class.

1. C S Smith and R J Forbes 'Metallurgy and Assaying' in Charles Singer *et al* A History of Technology 3 (Oxford 1957) 54.
2. Henry Cort 'Shingling, Welding and Manufacturing Iron and Steel' British Patent 1420 (1784). A letter from Black was published by Cort as a testimonial in Henry Cort *A Brief State of Facts relative to a New Method of Making Bar Iron* (1787) ('Remarks on Experiments made to Prove the Strength of Mr Cort's Iron').
3. Edinburgh University Library MS Gen 873 volume 3,f85 (letter from John Cologan to Black, 5 August 1788, "This friend wants Two of your Furnaces (Four à reverbere) the cost about 6 Guineas. I understand they are done at Carron").
4. In his lectures, Black referred to a reverberatory furnace for laboratory use: "In the *Journal de Physique,* 1776, there is a description of a Reverberatory Furnace, which seems very ingeniously fitted for a general laboratory furnace" (see Joseph Black *Lectures on the Elements of Chemistry . . . published by John Robison 1* (Edinburgh 1803) 326). This refers to a paper by Guyton de Morveau 'Description d'un nouveau Fourneau de Laboratoire' *Observations sur la Physique 8* 117 (August 1776) which is a discussion of a furnace originally described by Peter Macquer in 1758.
5. Black *op cit* (4) *2* plate III. This plate is clearly taken from Guyton's paper, *op cit* (4).
6. Edinburgh University Library MS Gen 270, List of Specimens.

vi-ix. Wollaston's Scales of Chemical Equivalents, 4 kinds. 1858.275.30

In a paper published in 1814, William Hyde Wollaston proposed his 'Synoptic Scale of Chemical Equivalents', an aid to calculation of combining quantities of chemical substances.[1] The concept was not a novel one, but by producing tables of figures (using the arbitary standard of oxygen=10) on logarithmic scales on a slide rule, he popularised the idea and the instrument was in vogue during the first half of the 19th century until its use declined for various reasons.[2] Modifications of Wollaston's rule were produced, some authors suggesting a different standard and increasing the number of substances represented on the scale. Thomas Charles Hope demonstrated a variety of scales to his classes. His manuscript List of Specimens (used by his assistant for preparing apparatus for demonstration at Hope's lectures) includes four: "Dr Wollastons Table Cheml Equivalents Dr Dewars Dr Reids — Prideaux".[3] These four may be those which once formed part of the Playfair Collection.

153

Scales based on the diagram in Wollaston's original paper seem to have been relatively common.[4] On the other hand, no examples of Dewar's, Reid's or Prideaux's tables appear to have survived. David Boswell Reid, Hope's assistant, published details of his scale in 1826.[5] The standard was based on hydrogen=1. John Prideaux's scale of 1830 was complex, including symbols for about 500 substances. It was based on oxygen=1 and was doubled, opening on hinges like a book.[6] 'Dr Dewars' is not easily traceable. The reference may be to a scale (unpublished?) by Henry Dewar MD, who in a paper of 1821 wrote:

> "Experimental analysis has reduced all bodies to 48 simple principles (including the most rare); and it is from the variety of the proportions and modes in which they are united, that the varieties of compounds, including vegetable and animal products arise."[7]

1. W H Wollaston 'A Synoptic Scale of Chemical Equivalents' *Phil Trans Royal Soc 104* 1 (1814).
2. Various forms of instrument are discussed by D C Goodman 'Wollaston and the Atomic Theory of Dalton' *Historical Studies in the Physical Sciences 1* 37 (1969). Michael Faraday *Chemical Manipulation* (London 1827) 551 warned on practical grounds that the instrument was not dependable: "It is almost impossible that the scales should be accurate, because of the extension and contraction of the paper when it is damped, and again dried, and the facility with which it yields to mechanical impressions."
3. Edinburgh University Library MS Gen 270, 'List of Specimens', under heading 'Chemical Action'.
4. There are three examples at the Science Museum, London (inv no 1932-578, see A Barclay *Handbook of the Collections Illustrating Pure Chemistry* (London 1937) 13), two examples at Harvard University (see David P Wheatland *The Apparatus of Science at Harvard 1765-1800* (Cambridge, Mass 1968) 173) and one example at the Museum of the History of Science, Oxford (see C R Hill *Museum of the History of Science Catalogue 1* (Oxford 1971) 42, item 292).
5. D B Reid *Directions for Using the Improved Sliding Scale of Chemical Equivalents* (Edinburgh 1826).
6. J Prideaux 'Continuation of the Table of Atomic Weights, and Notice of a new Scale of Equivalents' *The Philosophical Magazine 8* 423 (1830).
7. H Dewar 'The Influence of Chemical Laws on the Phenomena of Physiology' *Edinburgh Medical and Surgical Journal 17* 479 (1821).

x. **Alembic.** 1858.275.37
[Destroyed, Minute of Disposal Board, 4 April 1946]

This was not listed under "various forms of glass vessels used by Dr Joseph Black" (objects 40-64) so it may possibly be assumed that it was of clear rather than of green glass.

xi. **Flasks or receivers of various forms. (10 specimens).** 1858.275.39
[One specimen missing]

See notes for objects 43-51.

APPENDIX 2

Original entry in the manuscript Museum Register by George Wilson, Director of the Industrial Museum of Scotland. The numbers in italics following each entry refer to the numbering in this catalogue.

Presented by Professor Lyon Playfair   30 September 1858

Reg No		
275		Specimens from the Laboratory of the Edinburgh University, comprising Apparatus, instruments &c used by Professors Black, Hope and Gregory.
	1	Double barrelled Air Pump in the form in which it was first made by Hawksbee 1703-9. (*1*)
	2	Pneumatic Trough used by Dr Joseph Black in the collection of fixed air (Carbonic Acid) and other gases. (*2*)
	3	Balance used in his experiments by Dr Joseph Black, who was the first to employ the balance in chemical investigation. (*3*)
	4	Counterpoise (lead weight) for snow pan & lid (?) suspended from arm of Dr Black's balance wanting one of the scales. (*4*)
	5	Blow-pipe bellows invented by Sir Humphrey Davy. (*5*)
	6	Cage for placing ice in the focus of a Parabolic Mirror. (*6*)
	7	Pyrometer Ferguson's. (*7*)
	8	Thermometer formed of a spiral of brass. (*8*)
	9	Air Thermometer. (*9*)
	10	Differential Thermometers or Thermoscopes (3 specimens). (*10, 11, i*)
	10a	Double registering Thermometer (early form) by Adie, Edin^r (*ii*)
	11	Early form of the Voltaic Pile. (*12*)
	12	Cell of a Daniell's Battery. (*13*)
	13	Early form of Apparatus for decomposing water by the electrical current. (*14*)
	14	Stock of a pistol used by Professor Hope for shewing the mode of firing percussion caps when they were first introduced. (*iii*)
	15	Oxymuriate match box. The matches were tipped with chlorate of potash & sugar, which produced a light when dipped in the bottle containing sulphuric acid. This apparatus was in use before lucifer matches. (*15*)
	16	Another of a different form. (*16*)
	17	Argand Oil Lamp. (*17*)
	18	Experimental Oil Lamp. (*iv*)
	19	Lamp Furnace. (*18*)
	20	Sectional Model of a Reverberatory Furnace. (*v*)
	21	Old form of Laboratory Retort Stand. (*19*)
	22	Square Retort Stand enclosing a heating bath. (*20*)
	23	Support for glass tubes &c (*21 & 22*)
	24	Copper Still for careful distillations. (*23*)
	25	Liebig's Condensers. (*24 & 25*)
	26	Iron bottle for making Oxygen. (*26*)
	27	Model of Stool for Collecting gases in a pneumatic trough. (*27*)

Reg No		
275	28	A pair of brass reflectors, 12½ ins diameter. (*28 & 29*)
	29	A tin reflector, 12¼ ins diameter. (*30*)
	30	Wollaston's Scales of Chemical Equivalents, 4 kinds. (*vi - ix*)
	31	Tetragrammaton. (*31*)
	32	Iron Crucibles (3 specimens). (*32 - 35*)
	33	Arrangement for saturating water with Ammonia, hyxdrochloric acid &c. (*36*)
	34	Apparatus for transferring gases to a bladder. (*37*)
	35	Graduated glass receiver. (*38*)
	36	Bottle supposed to have been used by Dr Joseph Black in preparing Carbonic Acid. (*39*)
	37	Alembic. (*x*)

No 38 to No 45 are various forms of glass vessels used by Dr Joseph Black between 1766 & 1799. They are supposed to have been made at Leith.

	38	Solution Glasses figured in Dr Black's Lectures (3 specimens). (*40 - 42*)
	39	Flasks or receivers of various forms. (10 specimens). (*43 - 51, xi*)
	40	Retorts (3 specimens). (*52 - 54*)
	41	Lower portion of an Alembic, called a cucurbit by Dr Black. (*55*)
	42	Top or Capital of an Alembic. (*56*)
	43	Globular bottles (4 specimens). (*57 - 60*)
	44	Large bell shaped vessel. (*61*)
	45.	Two Electrophorus handles. (*62 & 63*)
	46	Arrangement for producing a light by igniting a jet of hydrogen gas by means of a spark from an electrophorus. (*64*)

APPENDIX 3

Apparatus and instruments which may have associations with the Edinburgh professors of chemistry in collections other than the Playfair Collection.

Andrew Plummer

1. Hydrostatic balance
(Private collection of Miss A Scott-Plummer)

A balance (with its original box) consisting of stand, beam (with swan-neck ends), weight pans, weights and a glass pear and bucket for measuring the specific gravity of liquids is still in the possession of a descendent of Andrew Plummer.[1] It is closely similar to a balance described in an article 'Hydrostatical-Ballance' attributed to Francis Hauksbee, published in 1710.[2] Plummer described using a hydrostatic balance in his investigation of the mineral waters of Moffat.[3]

A somewhat similar (though incomplete) balance, the property of St. John's College, Oxford, is on loan to the Museum of the History of Science, Oxford.[4]

1. R G W Anderson and A D C Simpson *Edinburgh & Medicine* (Edinburgh 1976) 35, item 130.
2. John Harris *Lexicon Technicum, or an Universal Dictionary of Arts and Sciences 2* (London 1710).
3. Andrew Plummer 'Experiments on the Medicinal Waters of Moffat' *Medical Essays and Observations 1* 82 (1733) ("The mineral Water tried by the hydrostatical Balance at the Fountain, and compared with the Water of a Rivulet near the Well, was found somewhat lighter than it; for the specific Gravity of the mineral Water was to that of the other Water as 838 to 840.")
4. C R Hill *Museum of the History of Science. Catalogue 1 Chemical Apparatus* (Oxford 1971) 48, item 323.

2-7. Glass bottles
(Private collection of Miss A Scott-Plummer)

Six small green glass bottles (three of phial shape, three squat) traditionally associated with Plummer. Three are closed with vellum and contain solid residues.

William Cullen

8. Mortar and pestle
(Royal College of Physicians of Edinburgh)

A somewhat crudely constructed bell-metal mortar and long pestle have traditionally been referred to as Cullen's though there is no documentary evidence to support this. The mortar has two loop handles and has the embossed initials 'RA'.[1]

1. John D Comrie *History of Scottish Medicine 1* (London 1932) 312 (showing an illustration of the mortar and pestle).

Joseph Black

9. Phial containing red powder
(Department of Chemistry, University of Edinburgh)

A specimen of what is probably ferric oxide has been sealed into a glass phial. It is likely that the powder was originally contained loose in a folded piece of paper, which accompanies the specimen, and was encapsulated at a later date. On the paper is written, in Black's hand: "Half a pinch of this Powder may be spread on your Strap & rubbed into it with the point of the finger once in 3 or 4 months". Other notes on the paper, written in three different hands, are: "Powder for Razor Strops prepared by Joseph Black M.D.", "Prof of Chemistry. The directions are in his handwriting" and "The writing is about 70 years old — (1787)".

10. Spirit thermometer
(Department of Technology, Royal Scottish Museum. Reg. no. 1869.11.1)

The thermometer is a crudely constructed spirit-in-glass type mounted on a wooden base, the manuscript scale on a paper strip stuck to the board running from 40° to $140^\circ F$. Its overall length is 632mm. It was presented to the Museum in 1869 by Matthew Forster Heddle, professor of chemistry at the University of St. Andrews from 1862 to 1884. Heddle stated that it had belonged to Black, though no further details of its provenance were suggested. It would seem strange that Black would have used this poorly made thermometer when he could call on the skill of his friend Alexander Wilson who constructed fine instruments (see chapter 1, ref. 30).

Thomas Charles Hope

11, 12. Samples of barium hydroxide and strontium hydroxide
(Department of Chemistry, University of Edinburgh)

These white precipitates in water are contained in sealed glass flasks, with handwritten labels 'baryta' and 'strontia'. Both are supported in japanned metal conical stands which have applied gold decoration depicting vines.[1] It is highly likely that the samples were prepared by Hope for display purposes, possibly shortly after his characterisation of strontium compounds in 1791-93 (see chapter 3).

1. John E Mackenzie 'The Chair of Chemistry in the University of Edinburgh in the XVIIIth and XIXth Centuries' *J Chem Ed* 12 503 (1935), fig 5.

13. Eudiometer
(Department of Chemistry, University of Edinburgh)

The eudiometer is constructed of thick glass and is graduated 0-110. It is fitted with two ground glass taps, one at the end of the tube, the other along one side, close to the end. Hope described an improved form of eudiometer in 1803.[1] Although this example does not conform to his description (which consisted of a graduated tube ground into a bottle which contained the absorbent) it would seem to be contemporary with it.

1. T C Hope 'Account of a simple Eudiometric Apparatus' *Nicholson's J* 6 61 *1803); 'Account of a Eudiometric Apparatus' *ibid* 210 and plate 12.

14. Chemical balance with weights

(Department of Chemistry, University of Edinburgh; currently (1978) on loan to the Department of Technology, Royal Scottish Museum, reg. no. 1970. L3)
Signed 'Robinson/38, Devonshire St./Portland Place, London.'

A fine analytical balance constructed by Thomas Charles Robinson, probably between 1829 and 1841. Robinson was an innovator in the construction of balances and made some of the finest examples of his time. This balance has a 10½ inch beam and these were advertised in 1829 at a price of 14 guineas.[1] A fitted drawer in the base contains two sets of grain weights in brass and platinum. It is possible that this balance was purchased by Hope.[2]

1. Advertisement in *Astronomische Nachrichten 7* 231 (1829).
2. John T Stock and D J Bryden 'A Robinson Balance by Adie & Son of Edinburgh' *Technology and Culture 13* 44 (1972).

14. **SELECT** balance with weights
Department of Chemistry, University of Edinburgh, currently (1978) on loan to the Department of Technology, Royal Scottish Museum, reg. no. 1979.1.31.
Signed 'Robinson/28, Devonshire St, Portland Place, London'.

A fine analytical balance constructed by Thomas Charles Robinson, probably between 1829 and 1841. Robinson 1.*Manufacturer* in the construction of balances and made some of the finest examples of his time. This balance has a 10½-inch beam and those were advertised in 1829 at a price of 14 guineas. A fitted drawer in the base contains two sets of grain weights in brass and platinum. It is possible that this balance was purchased by Hope.²

1. Advertisement in *Aitken's Gazette Advertising*, 2,251 (1829).
2. J. T. Stock and D. J. Bryden, "A Robinson Balance by Aitken, Son of Edinburgh, *Technology and Culture*, 13.44 (1972).

SELECT BIBLIOGRAPHY

1. Manuscripts

Aberdeen University Library
  MS 2206/24 'Account of Expenses for Apparatus &c ... by Dr William Gregory'
  MS 2206/25 William Gregory's 'Outlines of Lectures on Chemistry'

Birmingham Central Libraries
  Boulton and Watt Collection M3, James Watt's 'Ledger of Personal Accounts Jany 1764 to May 1769'
  Boulton and Watt Collection M3, 'Waste Book James Watt 1757'

British Library (Reference Division) Department of Manuscripts
  Sloane Collection MS 3216, f159 (letter from Archibald Pitcairne to Robert Gray)
  Add MS 42071 (letter from Joseph Black to Charles Greville)
  Add MS 52495 (lecture notes taken by a student of Joseph Black)
  National Reference Library of Science and Invention MS 40550 (lecture notes taken by a student of Joseph Black)

Chemical Society, London
  MS E.151.i 'Lectures on Chemistry by Joseph Black'

Edinburgh District Council Archives
  Town Council Record, 1675-1858
  Letter, 'Professor Black to B Steuart', 11 March 1780
  Report, John Walker to the Lord Provost and Town Council, 21 March 1780
  Plans of Alexander Monro's anatomy theatres.

Edinburgh University Library
  Minute Books of the Senatus Academicus, 1768-1844
  MSS Gen 873-75 (letters and other papers of Joseph Black)
  MSS Gen 268-72 (lecture notes and other papers of Thomas Charles Hope)
  MSS Dc.10.9-15 (Notes of lectures taken by Thomas Charles Hope when a student)
  MS Dc.2.76^{8*} (autobiographical fragment of Joseph Black)
  MS Dc.2.84 (lecture notes taken by a student of Joseph Black)
  MS Gen 48D (lecture notes taken by a student of Joseph Black)
  MS Gen 1959 'Minutes of the Professors of Medicine and Partners of the Chemical Elaboratory in Edinburgh'
  Architectural drawings of the College by William Henry Playfair, 1817

Glasgow University Archives
  Faculty Minute Books, 1747-1752
  University Minute Books, 1757-1795
  Inventories of apparatus in the chemistry laboratory, 1766-1773

Glasgow University Library
  MS Ferguson 41 (lecture notes taken by a student of Joseph Black)
  MS 2255/27 (letter of Walter Johnson to William Cullen)
  The Plan of A Course of Chemical Lectures... Glasgow MDCCXLVIII
  Cullen Papers

Library Company of Philadelphia
   MS Yi2. 1602.Q (lecture notes taken by a student of William Cullen 176-64)

National Library of Scotland
   MS 3533 (notes taken by Nathaniel Dimsdale of Joseph Black's lectures)
   MS 5077, f211 (letter of Joseph Black to Thomas Erskine)
   MS 5098, f49 (letter of Joseph Black to Thomas Erskine)

National Monuments Record of Scotland
   MS EDD/220/1 (scheme for rebuilding the College by Robert Morison)

Otago University Medical School Library, Dunedin
   MS M164 (notes taken by Alexander Monro *primus* at James Crawford's lectures)

Royal Botanic Garden of Edinburgh
   MS G.1265 (pressed botanical specimens of Robert Eliot)

Royal College of Physicians of Edinburgh
   MS M8/17-20 'Opera Chemica A. Plummer'

Royal College of Surgeons of Edinburgh
   Minute Books of the Incorporation of Barber-Surgeons, Incorporation of Surgeon-Apothecaries and Royal College of Surgeons of Edinburgh, 1657-1828
   MS bundle 92 (bill and receipt of Alexander Monteith, 1703)

Royal Infirmary of Edinburgh Archives
   Minute Books of the Infirmary Managers, 1752-1800

Royal Scottish Society of Arts
   MS Dep.230 (deposited in the National Library of Scotland) Minute Books of the Council and Committees, 1822-28

St Andrews University Library
   Muniment SM110. MB F37 (inventory of apparatus supplied by Alexander Allan to Robert Briggs)

Scottish Record Office
   Particular Register of Sasines, volumes 133 and 306
   MS RHP 9376 'Plan of the Present College of Edinburgh'
   Wills and inventories of Joseph Black, Thomas Charles Hope and William Gregory

Miss A Scott-Plummer
   'Book of Letters and Envoys Belonging to the Elaboratory'

Professor D C Simpson
   Letters from John Paterson to Robert Adam, 1790-91

Sir John Soane Museum, London
   Architectural drawings by Robert Adam, volume 28

Wedgwood Museum, Barlaston (on deposit in Keele University Library)
   Wd MS Mosley accumulation, letter Thomas Wedgwood to Alexander Chisholm, 11 January 1787
   MS 11-30437 (letter from Joseph Black to Josiah Wedgwood)
   MS 28-20002 (letter from John Wedgwood to Josiah Wedgwood)

Wellcome Institute for the History of Medicine
MS 2451 (notes taken by John Fullarton at James Crawford's lectures)

2. Unpublished theses

Cable, John A, 'Popular Lectures and Classes on Science in Scotland in the Eighteenth Century' M Ed thesis, University of Glasgow (1971)

Lawrence, Paul David, 'The Gregory Family: A Biographical and Bibliographic Study' Ph D Thesis, University of Aberdeen (1971)

McElroy, D D, 'The Literary Clubs and Societies of Eighteenth Century Scotland, and their Influence on the Literary Productions of the Period from 1700 to 1800' Ph D thesis, University of Edinburgh (1952)

3. Published works

Accum, Frederick, *Catalogue of Chemical Preprations, Apparatus and Instruments* (London 1805)

Accum, Frederick, *Descriptive Catalogue of the Instruments Employed in Chemistry* (London 1817)

Alison, William Pulteney, 'Account of the Life and Labour of Dr William Gregory' *Proc Royal Soc Edinburgh* 4 122 (1857-62)

Life of William Allen, with Selections from his Correspondence (London 1846)

Anderson, R G W and Simpson, A D C, *Edinburgh and Medicine* (Edinburgh 1976)

Badash, Lawrence, 'Joseph Priestley's Apparatus for Pneumatic Chemistry' *J Hist Med* 19 139 (1964)

Barclay, A, *Catalgue of the Collection in the Science Museum... Chemistry* (London 1927)

Black, Joseph, 'An Analysis of the Water of some Hot Springs in Iceland' *Trans Royal Soc Edinburgh* 3 95 (1794)

Black, Joseph, *De Humore Acido a Cibis Orto et Magnesia Alba* (Edinburgh 1754)

Black, Joseph, 'Experiments upon Magnesia Alba, Quicklime and some other Alcaline Substances' *Essays and Observations, Physical and Literary* 2 157 (1756)

Black, Joseph, *Lectures on the Elements of Chemistry... edited by John Robison* 2 volumes (Edinburgh 1803)

Brougham, Henry, Lord, *Lives of Philosophers of the Time of George III* (London and Glasgow 1855)

Brown, Leland A, *Early Philosophical Apparatus at Transylvania College...* (Lexington, Kentucky 1959)

Bryden, D J, *Scottish Scientific Instrument-Makers 1600-1900* (Edinburgh 1972)

The Caledonian Mercury (1724-1826)

Catalogue of the Special Loan Collection of Scientific Apparatus at the South Kensington Museum (London 1876)

Chaldecott, J A, *Handbook of the King George III Collection of Scientific Instruments* (London 1951)

Chaldecott, J A, *Heat and Cold Part II. Descriptive Catalogue* (London 1954)

Child, Ernest, *Tools of the Chemist* (New York 1940)

The Life of Sir Robert Christison, Bart. volume 1 (Edinburgh and London 1885)

Clow, Archibald and Clow, Nan L, *The Chemical Revolution* (London 1952)

Clow, Archibald and Clow, Nan L, 'Lord Dundonald' *The Economic History Review 12* 47 (1942)

Comrie, John D, *History of Scottish Medicine* 2 volumes (London 1932)

Coutts, James, *A History of the University of Glasgow from its Foundation in 1491 to 1909* (Glasgow 1909)

Cowen, David L, 'The Edinburgh Dispensatories' *The Papers of the Bibliographical Society of America 45* 85 (1951)

Cowen, David L, 'The Edinburgh Pharmacopoeia' in R G W Anderson and A D C Simpson (eds) *The Early Years of the Edinburgh Medical School* (Edinburgh 1976)

Crellin, J K and Scott, J R, *Glass and British Pharmacy 1600-1900* (London 1972)

Cresswell, C H, *The Royal College of Surgeons of Edinburgh 1505-1905* (Edinburgh 1926)

Cullen, William, 'Of the Cold produced by Evaporating Fluids and of Some Other Means of producing Cold' *Essays and Observations, Physical and Literara 2* 145 (1756)

Daumas, Maurice, *Scientific Instruments of the Seventeenth and Eighteenth Centuries and their Makers* (London 1972)

Donovan, A L, *Philosophical Chemistry in the Scottish Enlightenment* (Edinburgh 1975)

Duncan, Alexander, *Memorials of the Faculty of Physicians and Surgeons of Glasgow* (Glasgow 1896)

Edinburgh Evening Courant (1756-1826)

The Edinburgh New Dispensatory... by Gentlemen of the Faculty at Edinburgh (Edinburgh 1786)

An Enquiry into the General Effects of Heat; with Observations on the Theories of Heat and Mixture (London 1770)

Evidence, Oral and Documentary, taken by... the Commissioners... for Visiting The Universities of Scotland. Volume 1. University of Edinburgh (London 1837)

An Explanatory Dictionary of the Apparatus and Instruments of... Chemistry (London 1824)

Faraday, Michael, *Chemical Manipulation* (London 1827); *ibid* 3rd edition (London 1842)

Findlay, Alexander, *The Teaching of Chemistry in the Universities of Aberdeen* (Aberdeen 1935)

Fisher, George P, *Life of Benjamin Silliman M.D., LL.D.,* volume 1 (London 1866)

Fleming, J Arnold, *Scottish and Jacobite Glass* (Glasgow 1938)

Fletcher, Harold R and Brown, William H, *The Royal Botanic Garden Edinburgh 1670-1970* (Edinburgh 1970)

Forbes, R J, *A Short History of the Art of Distillation* (Leiden 1970)

Goodman, D C, 'Wollaston and the Atomic Theory of Dalton' *Historical Studies in the Physical Sciences 1* 37 (1969)

Graham, Thomas, *Elements of Chemistry* (London 1842)

Grant, Alexander, *The Story of the University of Edinburgh* 2 volumes (Edinburgh 1884)

Gray, James *History of the Royal Medical Society* (Edinburgh 1952)

Gray, Samuel, *The Operative Chemist* (London 1828)

Green, G Colman, 'William Gregory M.D., F.C.S.' *Nature 157* 467 (1946)

Gregory, William, *Letter to the Right Honourable George, Earl of Aberdeen, Kt. . . on the State of the Schools of Chemistry in the United Kingdom* (London 1842)

Gregory, William, *Letters to a Candid Inquirer, on Animal Magnetism* (Edinburgh 1851)

Gregory, William, 'Notes on the Purification and Properties of Chloroform' *Proc Royal Soc Edinburgh 2* 316 (1851)

Gregory, William, *Observations on the Proposed Appointment of a Teacher of Practical Chemistry in the University* (Edinburgh 1834)

Gregory, William, 'On a Process for Preparing Economically the Muriate of Morphia. With a letter from Dr Christison on its Employment in Medicine' *Edinburgh Med and Surgical J 35* 331 (1831)

Gregory, William, *Outlines of Chemistry for the Use of Students* London 1845)

Griffin, J J, *Chemical Handicraft* (London 1866); another edition (London 1877)

Griffin, J J, *Chemical Recreations* 8th edition (Glasgow 1838)

Guerlac, Henry, 'Black, Joseph' in Charles Coulston Gillispie (ed) *Dictionary of Scientific Biography* volume 2 (New York 1970)

Guerlac, Henry, 'Joseph Black and Fixed Air' *Isis 48* 124, 433 (1957)

Hill, C R, *Museum of the History of Science Catalogue 1: Chemical Apparatus* (Oxford 1971)

Holm, Knud E, 'Nolge Fyrtøjer fra 17- og 1800-tallet og deres Forudsaetninger med en Excurs til Kipp's Apparat' *Nationalmuseets Arbejdsmark* (Copenhagen 1971)

Hope, Thomas Charles, 'Account of a Mineral from Strontian and of a Peculiar Species of Earth which it Contains' *Trans Royal Soc Edinburgh 4* 3 (1798)

Hope, Thomas Charles, 'Account of a Simple Eudiometric Apparatus' *Nicholson's J 6* 61, 210 (1803)

Hope, Thomas Charles, 'Experiments and Observations upon the Contraction of Water by Heat at Low Temperatures' *Trans Royal Soc Edinburgh 5* 379 (1805)

Hope, Thomas Charles, 'Inquiry whether Sea-Water has its Maximum Density a few Degrees above its Freezing Point, as Pure Water has' *Trans Royal Soc Edinburgh 14* 242 (1840)

Horn, D B, *A Short History of the University of Edinburgh* (Edinburgh 1967)

Horn, D B, 'The Universities (Scotland) Act of 1858' *University of Edinburgh J 19* 169 (1959)

Kendall, James, 'The First Chemical Society, the First Chemical Journal, and the Chemical Revolution' *Proc Royal Soc Edinburgh 73* 346 (1952)

Kent, Andrew (ed), *An Eighteenth Century Lectureship in Chemistry* (Glasgow 1950)

King, W James, 'The Development of Electrical Technology in the 19th Century' *Contributions from the Museum of History and Technology Bulleting 228* (Washington DC 1963)

Lavoisier, A L, *Elements of Chemistry. . . translated by Robert Kerr* (Edinburgh 1790)

Leslie, John, *An Experimental Inquiry into the Nature and Propogation of Heat* (Edinburgh 1804)

Leslie, John, *A Short Account of Experiments and Instruments depending on the Relations of Air to Heat and Moisture* (Edinburgh 1813)

Lockemann, Georg, 'The Centenary of the Bunsen Burner' *J Chem Ed 33* 20 (1956)

Macbride, David, *Experimental Essays* (London 1764)

McKie, Douglas (ed), *Notes from Dr Black's Lectures on Chemistry 1767/8* (Wilmslow, Cheshire 1966)

McKie, Douglas and Kennedy, David, 'On some letters of Joseph Black and Others' *Annals of Science 16* 129 (1962)

Martine, George, *Essays and Observations on the Construction and Graduation of Thermometers. . . New Edition* (Edinburgh 1792)

Middleton, W E Knowles, *A History of the Thermometer* (Baltimore 1966)

Morgan, A (ed), *Charters, Statutes and Acts of the Town Council and Senatus* (Edinburgh 1937)

Morrell, J B, 'The Edinburgh Town Council and its University, 1717-1766' in R G W Anderson and A D C Simpson (eds), *The Early Years of the Edinburgh Medical School* (Edinburgh 1976)

Morrell, J B, 'Practical Chemistry in the University of Edinburgh, 1799-1843' *Ambix 16* 66 (1969)

Oddy, Revel, 'Scottish Glass Houses' *The Circle of Glass Collectors Paper 151* (1966)

Oddy, Revel, 'Some Chemical Apparatus blown by hand in the late 18th to early 19th century' *Annales du 5e Congrès de l'Association Internationale pour l'Histoire du Verre* (Liège 1972)

Partington, J R, *A History of Chemistry* volumes 2, 3 and 4 (Ldondon 1961-64)

Pictet, M A, *An Essay on Fire* (London 1791)

Playfair, Lyon, *A Century of Chemistry in the University of Edinburgh: being the Introductory Lecture to the Course of Chemistry in 1858* (Edinburgh 1858)

Playfair, William Henry, *Report Respecting the Mode of Completing the New Buildings for the College of Edinburgh* (Edinburgh 1816)

Plummer, Andrew, 'Experiments on the Medicinal Waters of Moffat' *Medical Essays and Observations 1* 82 (1733)

Plummer, Andrew, 'Remarks on chemical Solutions and Precipitations' *Essays and Observations, Physical and Literary 1* 284 (1754); 'Experiments on Neutral Salts, compounded of different acid liquors, and alcaline salts, fixt and volatile' *ibid* 315

Porter, Arthur L, *The Chemistry of the Arts* 2 volumes (Philadelphia 1830)

Ramsay, William, *The Life and Letters of Joseph Black, MD* (London 1918)

Reid, David Boswell, *Elements of Chemistry* 3rd edition (Edinburgh 1839)

Reid, David Boswell, *A Memorial to the Patrons of the University on the Present State of Practical Chemistry* (Edinburgh 1833)

Reid, David Boswell, *Remarks on the Present State of Practical Chemistry and Pharmacy* (Edinburgh 1838)

Reid, Wemyss, *Memoirs and Correspondence of Lyon Playfair* (London 1899)

Fourth and *Sixth Report of the Science and Art Department to the Committee of Council on Education* (London 1857 and 1859)

Reuss, August Christian, *Beschreibung eines neues Chemischen Ofens* (Leipzig 1782)

Ritchie, R Peel, *The Early Days of Royall Colledge of Phisitians, Edinburgh* (Edinburgh 1899)

Robinson, Eric and McKie, Douglas, *Partners in Science* (London 1970)

Schrøder, Michael, *The Argand Burner* (Odense 1969)

Speter, Max, 'Geschichte der Erfindung des Liebigischen Kühlapparates' *Chemiker Zeitung 32* 3 (1908)

Stock, John, *Development of the Chemical Balance* (London 1969)

Swinbank, P, 'James Watt and his Shop' *Glasgow University Gazette 59* 4 (1969)

Taylor, E G R, *The Mathematical Practitioners of Hanoverian England* (Cambridge 1966)

Taylor, F Sherwood, 'The Evolution of the Still' *Annals of Science 5* 185 (1945)

Testimonials in Favour of William Gregory, M.D., F.R.S.E. (Edinburgh 1853)

Testimonials Regarding Dr D. B. Reids Qualifications as a Lecturer on Chemistry and as a Teacher of Practical Chemistry (Edinburgh 1833)

Thomson, John, *An Account of the Life. . . of William Cullen* 2 volumes (Edinburgh and London 1859)

Thomson, Thomas, 'Dr Joseph Black' in *The Edinburgh Encyclopaedia 3* 549 (Edinburgh 1830)

Traill, Thomas Stewart, 'Memoir of Dr Thomas Charles Hope' *Trans Royal Soc Edinburgh 16* 419 (1849)

Turner, A Logan, *Story of a Great Hospital: The Royal Infirmary of Edinburgh 1729-1929* (Edinburgh 1937)

Turner, G L'E and Levere, T H, *Martinus van Marum: Life and Work Volume IV. Van Marum's Scientific Instruments* (Leyden 1973)

Underwood, E Ashworth, *Boerhaave's Men at Leyden and After* (Edinburgh 1977)

Walton, P M Eaves, 'The Early Years of the Infirmary' in R G W Anderson and A D C Simpson (eds) *The Early Years of the Edinburgh Medical School* (Edinburgh 1976)

Wheatland, David P, *The Apparatus of Science at Harvard 1765-1800* (Cambridge, Mass 1968)

Wightman, William P D, 'William Cullen and the Teaching of Chemistry' *Annals of Science 12* 154, 192 (1956)

Williams, C Greville, *A Handbook of Chemical Manipulation* (London 1857)

Wilson, Jessie Aitken, *Memoir of George Wilson* (Edinburgh 1860)

Wright-St Clair, R E, *Doctors Monro* (London 1964)

INDEX OF PERSONAL NAMES

Those names listed in italics appear only in the Catalogue Section

Adam, Robert (1728 – 1792)	22, 23, 37
Adam, William (1689 – 1748)	10
Accum, Friedrich Christian (1769 – 1838)	78, 97, 102, 105, 120
Adie, Alexander (1774 – 1858)	40, 60, 89, 151
Adie, Richard (1810 – 1881)	151
Alison, William Pulteney (1790 – 1859)	53
Allan, Alexander (fl.1804 – 1835)	60
Allen, William (1770 – 1843)	39, 40, 58, 89
Alston, Charles (1683 – 1760)	5, 8, 13, 19
Apjohn, James (1796 – 1886)	48
Anderson, James (1739 – 1808)	13
Anderson, John Wilson (fl.1823 – fl.1836)	38, 39
Argand, François-Pierre Ami (1750 – 1803)	101, 102
Ash, Edward (fl.1786 – 1829)	36
Astley, Joseph (fl.1818 – 1831)	39
Astley, Thomas (fl.1833 – 1850)	39
Babington, William (1756 – 1833)	96
Baker, George (1722 – 1809)	35
Banks, Joseph (1744 – 1820)	35, 91
Bannerman, James (fl.1792 – 1838)	48
Barrett, Francis (fl.1801 – fl.1815)	124
Becher, Johann Joachim (1635 – ?1682)	11, 24
Beddoes, Thomas (1769 – 1808)	92, 115, 116
Bennett, John Hughes (1812 – 1875)	53
Berzelius, Jöns Jacob (1779 – 1848)	48, 50, 102, 105
Black, Alexander (1729 – 1813)	24, 143
Black, Joseph (1728 – 1799)	ix, 1, 10, 12, 13, 19–27, 35, 36, 40, 41, 42, 57, 58, 59, 60, 65, 66, 69, 71, 73, 74, 76, 77, 78, 81, 86, 92, 101, 102, 110, 111, 115, 116, 120, 125, 127, 128, 132, 134, 135, 138, 142, 143, 144, 145, 153, 154, 158
Blagden, Charles (1748 – 1820)	35
Bloxam, Thomas (fl.1855 – fl.1871)	142
Boerhaave, Herman (1668 – 1738)	4, 8, 21
Bonnar, Thomas (fl.1795 – ?1832)	37
Borthwick, James (1615 – 1676)	3
Borthwick, William (fl.1653 – 1689)	3
Boswell, James (1740–1795)	26
Breguet, Abraham-Louis (1747 – 1823)	84, 85
Boyle, Robert (1627 – 1691)	67
Brewin, T	92
Brougham, Henry, Lord Brougham and Vaux (1778 – 1868)	27
Brown, David Rennie (1808 – 1875)	39
Brown, Samuel (1817 – 1856)	50
Brown, William (1733 – 1797)	77

Brownrigg, William (1711 – 1800)	118
Bucquet, Jean Baptiste Michel (1746 – 1780)	127
Bunsen, Robert (1811 – 1899)	50
Button, Charles (fl.1846 – fl.1857)	113
Carlisle, Anthony (1768 – 1840)	92, 96
Cary, William (1759 – 1825)	89
Carpue, Joseph Constantine (1764–1846)	92
Carrick, John (fl.1746 – 1750)	10
Carstares, William (1649 – 1715)	4
Cavallo, Tiberius (1749 – 1809)	128
Cave, Joseph (fl.1713 – fl.1728)	5
Cavendish, Henry (1731 – 1810)	35, 73, 132
Children, John George (1777 – 1852)	47, 95
Christison, Robert (1797 – 1882)	36, 38, 39, 53
Clark, Thomas (1801–1867)	48
Cleghorn, Robert (?1755 – 1821)	35
Clow, James (fl.1731 – 1788)	20
Cochrane, Archibald, Earl of Dundonald (1749 – 1831)	23, 24
Cochrane, Thomas (fl.1766 – 1782)	24
Cockburn, Henry Thomas, Lord Cockburn (1779 – 1854)	42, 53
Collinson, Peter (1694 – 1768)	144
Cort, Henry (1740 – 1800)	24, 153
Crawford, Adair (?1748 – 1795)	35
Crawford, James (? – ?1732)	4, 5, 58
Crichton, James (fl.1775 – fl.1835)	84
Cronstedt, Axel Frederic (1702 – 1765)	77, 78
Cruikshank, William (? – 1811)	35, 92, 96
Cullen, William (1710 – 1790)	1, 10–14, 19, 20, 21, 22, 24, 26, 27, 35, 58, 65, 66, 69, 143, 157
Cunningham, Patrick (fl.1661 – 1686)	3
Dalrymple, John (1726 – 1810)	24
Dalzell, Allen (fl.1850 – fl.1866)	53
Damion, John (fl.1501 – 1513)	142
Daniell, John Frederick (1790 – 1845)	94, 97
D'Arcet, Jean (1725 – 1801)	107
Darwin, Charles (1809 – 1882)	36
Davy, Humphry (1778 – 1829)	77, 78, 92, 95
De Boynes, Pierre–Etienne (1718 – 1783)	113
Defoe, Daniel (1661 – 1731)	142
Desaguliers, John Theophilus (1683 – 1744)	68
Deuchar, John (fl.1815 – 1833)	60
Dewar, Henry (fl.1797 – 1822)	153, 154
Dick, Robert (1722 – 1757)	19
Döbereiner, Johann Wolfgang (1780 – 1849)	149
Drummond, George (1687 – 1766)	5, 8, 10, 13, 58
Duncan *senior,* Andrew (1744 – 1828)	35
Duncan *junior,* Andrew (1773 – 1832)	37
Dundas, Henry, Viscount Melville (1742 – 1811)	22
Dunlop, Alexander (fl.1744 – 1750)	10
Dunn, John (fl.1824 – 1841)	40, 47, 60, 82, 83, 85

Ehrmann, Friedrich Ludwig (1741 – 1800)	148
Eliot, Robert (fl.1688 – 1714)	4
Elliot, Thomas (fl.1740 – 1751)	10
Engestroem, Gustav von (1738 – 1813)	78
Erskine, Thomas A, Earl of Kellie (1732 – 1781)	24

Faraday, Michael (1791 – 1867)	47, 97, 113, 125
Farey, John (fl.1769)	22
Ferguson, Adam (1723 – 1816)	19
Fleming, John (1785 – 1857)	50
Forbes, James (1809 – 1868)	36, 48
Forbes, Patrick (1776 – 1847)	49, 59
Forsyth, Alexander John (1768 – 1843)	152
Fortin, Nicholas (1750 – 1831)	73
Fothergill, John (1712 – 1780)	9
Fourcroy, Antoine Francois de (1755 – 1809)	132
Frankland, Edward (1825 – 1899)	1
Franklin, Benjamin (1706 – 1790)	144
Frotheringham, Samuel (fl.1747)	82
Fullerton, John (fl.1714)	4
Fyfe, Andrew (1792 – 1861)	38, 39, 41, 50

Gadolin, Jacob (1719 – 1802)	113
Gadolin, Johan (1760 – 1852)	113
Gahn, Johan Gottlieb (1745 – 1818)	105
Garbett, Samuel (1717 – 1805)	24
Gascoigne, Charles (?1738 – 1806)	24
Gay-Lussac, Joseph Louis (1778 – 1850)	105
Geddes, Archibald (fl.1786 – fl.1808)	24, 26, 143
Geddes, William (fl.1794 – fl.1818)	24, 143
Geoffroy, Etienne Francois (1672 – 1731)	11
Gmelin, Leopold (1788 – 1853)	60, 106
George III, King of Great Britain and Ireland (1738 – 1820)	23, 35
Gordon, George Hamilton, Earl of Aberdeen (1784 – 1860)	48, 49
Graham, George (1673 – 1751)	82
Graham, Robert (fl.1799)	24
Graham, Thomas (1805 – 1869)	48, 50, 57, 113
Gregory, James (1638 – 1675)	47
Gregory, James (1753 – 1821)	35, 47
Gregory, John (1724 – 1773)	13, 47
Gregory, Olinthus (1774 – 1841)	94
Gregory, William (1803 – 1858)	1, 2, 39, 47–53, 57, 58, 59, 60, 65, 66, 83, 110, 123, 144, 151
Griffin, John Joseph (1802 – 1877)	60, 61, 105, 106, 109, 113, 118, 125
Griffin, Richard Thomas (?1789 – 1832)	60

Hales, Stephen (1677 – 1761)	71
Halket, James (?1654 – 1711)	4
Hall, James (1761 – 1832)	35
Hamilton, Robert (1714 – 1756)	10, 19
Hare, Robert (1781 – 1858)	40, 94, 95

Harris, John (1667 – 1719)	73
Harrison, John (1693 – 1776)	82
Harrison, Thomas (fl.ca.1775)	73
Hassenfratz, Jean Henri (1755 – 1827)	127
Hauksbee the elder, Francis (?1666 – 1713)	67–69, 157
Hauksbee the younger, Francis (1688 – 1763)	14, 69
Hay, George, Earl of Kinnoul (1572 – 1634)	142
Heddle, Matthew Forster (1828 – 1897)	21, 158
Henry, William (1774 – 1835)	96, 132
Heriot, Thomas (fl.1759 – fl.1786)	22
Herschel, William (1738 – 1822)	35
Hill, Ninian (1728/9 – 1790)	20, 21
Home, Francis (1719 – 1813)	12, 24, 35, 58
Home, Henry, Lord Kames (1696 – 1782)	12
Hooke, Robert (1635 – 1703)	67
Hope, James, Earl of Hopetoun (1741 – 1816)	24
Hope, John, Earl of Hopetoun (1704 – 1781)	24
Hope, John (1725 – 1786)	35
Hope, Thomas Charles (1766 – 1844)	1, 2, 24, 26, 27, 35–42, 47, 48, 50, 53, 57, 58, 60, 65, 66, 78, 81, 82, 84, 86, 89, 92, 100, 104, 110, 111, 113, 116, 121, 128, 143, 144, 149, 151, 152, 153, 154, 158
Hopson, Charles Rivington (1744 – 1796)	78
Hutcheson, Francis (fl.1748 – fl.1758)	20
Innes, John (1696 – 1733)	5, 8, 9, 58, 142
Irvine, William (1743 – 1787)	21, 22, 35, 108, 143
Irving, Ralph (1760 – 1795)	111
James IV, King of Scotland (1473 – 1513)	142
Johnson, Walter (fl.1747)	11, 14, 143
Johnstoun, Alexander (fl.1733 – fl.1777)	143
Johnstoun, John (1684 – 1762)	10
Keir, James (1735 – 1820)	26
Keir, Peter (fl.1785 – fl.1787)	102
Kemp, Alexander (1822 – 1854)	52
Kemp, Kenneth Treasurer (1805 – 1842)	39, 60
Kincaid, Alexander (ca.1753 – fl.1794)	13, 22
Kincaid, Thomas (1618/9 – 1691)	3
King, James (1800 – 1857)	151
Lambert, Johann Heinrich (1728 – 1777)	84
Laurie, Gilbert (fl.1742 – fl.1787)	9
Laurie, John (fl.1758 – fl.1794)	10
Lavoissier, Antoine-Laurent (1743 – 1794)	1, 26, 35, 36, 40, 73, 113, 127, 130, 144
Leechman, William (1706 – 1785)	20
Leslie, John (1766 – 1832)	81, 86, 89, 90, 120

Lewis, William (1714 – 1781)	111
Liebig, Justus (1803 – 1873)	48, 49, 50, 52, 57, 83, 113, 144
Lind, James (1738 – 1812)	127, 128
Lorenz, Richard (fl.1807)	148
Macbride, David (1726 – 1778)	20
Macfarlan, John Fletcher (fl.1780 – 1830)	39, 48, 60
Macie, James Louis (1765 – 1825)	74
Maclaurin, Colin (1698 – 1746)	9
Macquer, Pierre Joseph (1718 – 1784)	153
Magellan, John Hyacinthe de (1722 – 1790)	77, 78, 84, 113
Mariotte, Edme (?1620 – 1684)	120
Martin, Benjamin (1704 – 1782)	69
Martine, George (1702 – 1741)	11, 21
Marum, Martinus van (1750 – 1837)	1, 81, 144
Maydwell, Glisson (fl.1726 – 1742)	143
Mayer, Matthew James (fl.1811)	148, 149
Mayow, John (1640 – 1679)	71
Mégnie, Pierre Bernard (?1751 – 1807)	73
Miller, John (fl.1769 – 1822)	60, 89, 151
Miller, Thomas, Lord Glenlee (1717 – 1789)	20
Monro *primus*, Alexander (1697 – 1767)	4, 5, 8, 13
Monro *secundus*, Alexander (1733 – 1817)	10, 22, 35
Monro *tertius*, Alexander (1773 – 1859)	37
Monro, John (1670 – 1740)	5, 8
Monteith, Alexander (fl.1684 – 1713)	3, 4
Mort, Jacobus Le (1650 – 1718)	4
Mortimer, Cromwell (?1698 – 1752)	82
Murchiston, Roderick Impey (1792 – 1871)	41
Musschenbroek, Petrus van (1692 – 1761)	82
Neville, Sylas (1741 – 1840)	27
Nicholson, William (1753 – 1815)	40, 91, 92, 96
Noad, Henry M (fl.1838 – fl.1877)	95
Nollet, Jean Antoine (1700 – 1770)	120
Nooth, John Mervin (fl.1763 – 1828)	78, 143
Ostwald, Wilhelm (1853 – 1932)	97
Papin, Denis (1647 – 1714)	67, 68
Parkes, Samuel (1761 – 1825)	71
Paterson, John (fl.1776 – 1832)	23
Perceval, Robert (1756 – 1839)	103, 104
Pictet, Marc Auguste (1752 – 1825)	80, 81, 121
Pitcairne, Archibald (1652 – 1713)	3, 4
Playfair, John (1748 – 1819)	81
Playfair, Lyon (1818 – 1898)	ix, 1, 52, 53, 57, 59, 65, 73, 74, 76, 113, 155

Playfair, William Henry (1789 – 1857)	37
Plummer, Andrew (1698 – 1756)	5, 8–10, 11, 12, 13, 20, 22, 26, 58, 65, 66, 69, 73, 142, 157
Pott, Johann Heinrich (1692 – 1777)	11
Preston, Charles (1660 – 1711)	5
Preston, George (?1665 – 1749)	5
Prideaux, John (fl.1823 – fl.1852)	153, 154
Priestley, Joseph (1733 – 1804)	130
Ramsden, Jesse (1735 – 1800)	73, 74
Reichenbach, Carl von (1788 – 1869)	52, 123
Reid, David Boswell (1805 – 1863)	38, 39, 40, 47, 48, 50, 60, 95, 113, 144, 153, 154
Reid, Robert (1774 – 1856)	37
Reid, Thomas (1710 – 1796)	24
Rinman, Sven (1720 – ?1794)	77
Ritter, Johann Wilhelm (1776 – 1810)	96
Robertson, E G (fl.1800)	97
Robertson, William (1721 – 1793)	22
Robinson, Thomas Charles (1792 – 1841)	60, 159
Robiquet, Pierre Jean (1780 – 1840)	47, 48
Robison, John (1739 – 1805)	19, 20, 21, 22, 26, 35, 71, 92, 101, 102, 111, 143, 144, 145, 153
Rotheram, John (?1750 – 1804)	27
Russell, James (fl.1739 – 1773)	19
Rutherford, Daniel (1749 – 1819)	27, 35, 151
Rutherford, John (1695 – 1779)	5, 8, 9, 10, 13, 58, 142
Rutherford, John (fl.1790 – fl.1798)	151
St Clair, Andrew (1697 – 1760)	5, 8, 9, 10, 13, 58, 142
Santorio, Santorio (1561 – 1636)	86
Saussure, Horace Benedict de (1740 – 1799)	121
Scott, James (fl.1739 – fl.1755)	9, 10, 13, 22, 58
Sefstroem, Nils Gabriel (1787 – 1845)	105
Shaw, Peter (1694 – 1763)	14
Sibbald, John (fl.1764 – 1818)	24, 26, 60
Sibbald, Robert (1641 – 1722)	4
Silliman, Benjamin (1779 – 1864)	36, 37, 38, 40, 144
Simon, Paul Louis (1767 – 1815)	97
Simpson, James Young (1811 – 1870)	52
Sinclair, Catherine (1800 – 1864)	53
Six, James (fl.1782 – 1793)	151
Smith, Henry (1810 – 1865)	39
Smith, Thomas (1807 – 1893)	39
Stahl, Georg Ernst (1660 – 1734)	11
Stevenson, Alexander (fl.1749 – 1791)	35
Stewart, Dugald (1753 – 1828)	19, 81
Stewart, John (fl.1742 – fl.1759)	69
Sturgeon, William (1783 – 1850)	94, 95
Sutherland, James (fl.1670 – 1719)	4, 5

Telford, Thomas (1757 – 1834)	41
Thompson, Benjamin, Count Rumford (1753 – 1814)	40, 81, 89
Thomson, Allen (1809 – 1884)	52
Thomson, Thomas (1773 – 1852)	38, 39, 50, 60, 120, 132
Traill, Thomas Stewart (1781 – 1862)	39, 41, 42, 50, 58
Trithemius, Johannes (1462 – 1516)	124
Turner, Edward (1798 – 1837)	39, 47, 48, 50
Turner, Wilson G (fl.1842 – fl.1846)	50
Volta, Alessandro (?1745 – 1827)	91, 92, 96
Vream, William (fl.1710 – fl.1717)	68
Walker, John (1731 – 1803)	22, 35
Walker, John (?1781 – 1859)	100
Watt, James (1736 – 1819)	20, 21, 22, 23, 24, 26, 27, 71, 92, 101, 102, 111, 115, 116, 120
Webster, Charles (fl.1773 – 1795)	102, 111
Wedgwood, Josiah (1730 – 1795)	26, 27, 82
Wedgwood, Thomas (1771 – 1805)	27, 115, 116
Weigel, Christian Ehrenfeld (1748 – 1831)	113
Whewell, William (1794 – 1866)	97
Whytt, Robert (1714 – 1766)	10, 13, 19
Wilkinson, Charles Henry (fl.1799 – fl.1830)	92
Williams, Charles H Greville (1829 – 1910)	105
Wilson, Alexander (1714 – 1786)	20, 21, 22, 26, 69, 158
Wilson, George (1818 – 1859)	ix, x, 1, 39, 41, 53, 57, 58, 65, 71, 74, 76, 142, 155
Wilson, Patrick (1743 – 1811)	22
Wöhler, Friedrich (1800 – 1882)	50
Wollaston, William Hyde (1766 – 1828)	95, 153, 154
Woulfe, Peter (?1727 – ?1805)	108, 127
Zimmerman, Carl Friedrich (fl.1739 – fl.1746)	77